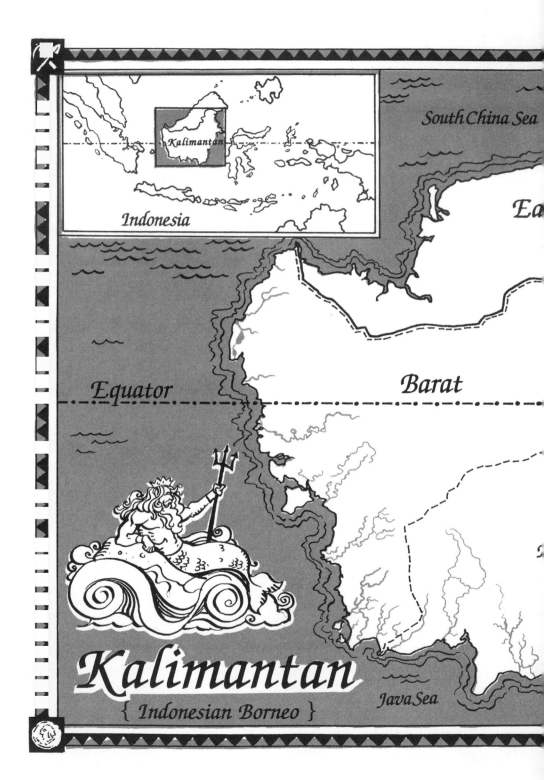

South China Sea

Kalimantan

Indonesia

Ea

Equator

Barat

Java Sea

Kalimantan
{ Indonesian Borneo }

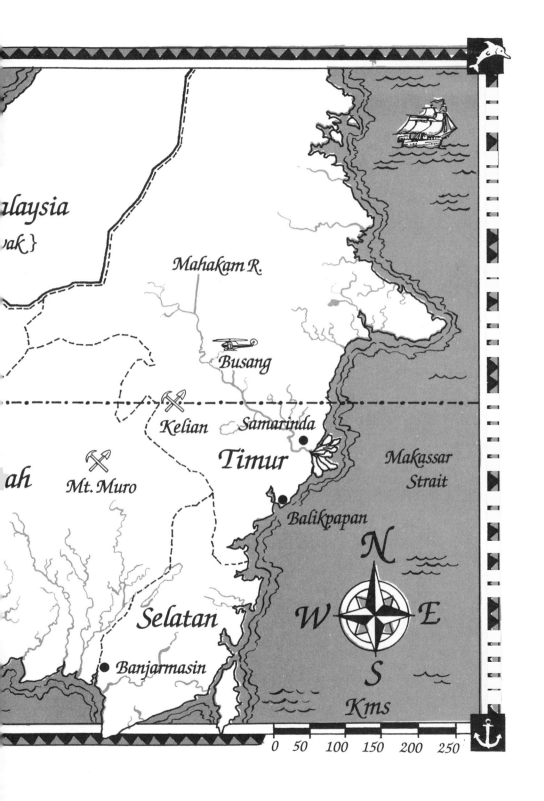

FOOLS' GOLD

FOOLS'

Brian Hutchinson

GOLD

The Making of a
Global Market Fraud

Alfred A. Knopf Canada

PUBLISHED BY ALFRED A. KNOPF CANADA

Canadian Cataloguing in Publication Data

Hutchinson, Brian, 1965
    Fools' gold: the making of a global market fraud

ISBN 0-676-97098-2

1. Bre-X (Firm)   2. Gold mines and mining - Indonesia   3. Fraud.

I. Title

HD9536.I544B78 1997   338.2'741'09598   C97-931522-0

Cover design: Spencer Francey Peters
Page design: Gordon Robertson
Map: Paul McCusker

First Edition

Printed and bound in the United States of America

CONTENTS

ACKNOWLEDGEMENTS

FOOLS' GOLD was inspired by an article I wrote for *Canadian Business* magazine early in 1997. I am grateful to the magazine's editor, Art Johnson, who assigned me to what developed into an unforgettable, fourteen-month adventure. Many thanks to Ian McGugan, who helped set the tone, and Sean Silcoff, whose investigative work cast light on several of this book's characters. I also received valuable support from colleagues Bruce Headlam, Brian Banks, Peter Shawn Taylor, Donna Braggins, Pat Ireland and Miguel Rakiewicz.

While it's a given that this book will be condemned by Indonesia's ruling authority, I offer my best wishes to the Indonesian people, particularly those who put themselves at risk while explaining how their systems work. I am also grateful to those Canadian mining executives and investment advisors who spoke honestly about their professions.

Other individuals, organizations and companies that supplied me with valuable information include Arif Arryman, Ngang Bilung, Marian Botsford-Fraser, Kevin Evans, Sarkarni Gambi, Sean Gordon, Kurdiansyah, Wahid Rahmanto, William Shields, Tom Soulsby, the Indonesian Forum for the Environment, the Indonesian Mining Association, Indonesia's Department of Mines and Energy, Econit, East Timor Alert Network, the Institute for Policy Research and Advocacy, Barrick Gold Corporation, Deloitte and Touche Inc., PT

Freeport Indonesia, Inco Limited, Indomin Resources Limited, Placer Dome Inc., Price Waterhouse Limited, Strathcona Mineral Services Limited, the Alberta Securities Commission, the Ontario Securities Commission, and the Toronto Stock Exchange.

Dianna Symonds and Isabel Vincent offered heartfelt advice as I wrote this book and kept me aware of the end. Douglas Thomson's excellent research added depth to the project, and David Kilgour was a thorough and thoughtful editor. Wendy Thomas did a remarkable job with the copy editing. My hat is off to the people at Alfred A. Knopf Canada, who always made me feel welcome; I am especially grateful to Louise Dennys and Susan Burns for their guidance and friendship. Anne McDermid, my literary agent, was indefatigable and a source of sound creative advice. Ken Whyte and Mark Stevenson have been particularly supportive since helping launch my writing career in 1989.

To my family and friends, thanks for the encouragement. To Judy Cheung, my inspiration and my compass, I dedicate this book, with love.

WHAT IS PAST IS PROLOGUE

Such would be the delusion, that when the evil day came, as come it would, the people would start up, as from a dream, and ask themselves if these things could have been true.

    – Charles Mackay, *Extraordinary Popular Delusions and the Madness of Crowds*

A **STOCK PROMOTER** sits in his cluttered office. He's sweating profusely and shifting his eyes as a pair of alert investment professionals grill him about his company's performance. They demand to see every item in his ledger and want to scrutinize every asset. They evaluate his character. He's failing their test. "I know it looks bad," blurts the promoter, "but I have an explanation." The investment professionals glance at each other, left eyebrows raised. *This guy's a dud*, their expressions say. *He's not getting our clients' money, case closed. Time to look somewhere else.*

That's how the game works, according to countless print advertisements and flashy television spots. *We dig deeper.* The message is delivered with mind-numbing frequency at the start of every new year, when Canadians are asked to choose between Brand X mutual fund and Brand Y retirement plan. We obey, shelling out more of our savings, hoping to tap into an unparalleled spurt of growth in the booming equities market.

And then this: the greatest mining swindle in history, manufactured in Canada. The rise and fall of Bre-X Minerals Ltd., the tiny exploration company that seduced the sharpest investment professionals in the business, was a grotesque example of greed and deception. But it was perfectly symptomatic of a disease that swept, unchecked, through an entire industry. Bre-X claimed to have found a monstrous deposit of gold in distant Indonesia, and tapped into a Canadian-led gold rush, an investment bonanza that sucked billions of dollars from trusting investors and siphoned the money to cynical insiders and stock market sharpies. The incident involved promoters, analysts, regulators and media, embroiled in a silent conspiracy of consent.

The scam was revealed, the bubble burst. The flow of cash pouring into Canada's mining industry slowed to a trickle, and companies big and small shut down their operations, laid off staff, or imploded. Now the bankruptcy trustees and the class-action lawyers pick their way through the ruins. Blame is shunted back and forth, but there is no accountability, not yet.

As I began to explore this story, travelling across Canada, to Australia, Southeast Asia, and into the Borneo rainforest, I realized that there was far more to it than one disingenuous company and a sensational fraud. This is a complex morality play, in which thousands of unsuspecting Canadians played a part, and involving hundreds of characters working in concert and at cross purposes. It is about corrupt political regimes and financial institutions, carefree regulatory bodies, and irresponsible media. It is about the tarnishing of Canada's mining industry, once the greatest in the world, and of our willingness to exploit indigenous peoples and despoil their environment for profit. It is about the very human desire to believe, against all reason, in buried treasure. It is about lies and broken trust. Ordinary investors — neighbours, relatives, colleagues — put their faith in a system and they were betrayed again and again. Some never recovered, and took their own lives. Others struggled on, their trust in the instruments of commerce shattered.

Stock promoters and gold analysts and the money men on Bay Street claim that Bre-X was an anomaly. Bre-X was not a fluke. It was inevitable. We should have seen it coming.

In 1841, a British lawyer named Charles Mackay wrote a remarkable book called *Extraordinary Popular Delusions and the Madness of Crowds*. It's a delicious examination of mass delirium and assorted mania. Along with chapters on witch burning and religious persecutions, Mackay's book explains the logic, or lack thereof, behind the South Sea Bubble, a telling example of investment mania that turned England upside down during the eighteenth century. In the wake of an over-subscribed public venture to import gold and silver from the New World, a "speculative frenzy" was unleashed on the streets of London, fuelled by incredible promises of wealth. Thousands of ordinary citizens invested in dozens of ridiculous money-raising schemes that had no hope of paying off. Fraud was so rampant that the "sensible men" in Parliament denounced eighty-six enterprises as public nuisances and declared them illegal. Among the banned bubbles were missions to trade human hair, to improve the art of making soap, and to improve the quality of English gardens. "The most absurd and preposterous of all," Mackay wrote, "and which shewed, more completely than any other, the utter madness of people, was one started by an unknown adventurer, entitled, *'A company for carrying on an undertaking of great advantage, but nobody to know what it is.'*" The chicanery was outlawed, but it didn't disappear. It simply went underground, with the rest of them.

After reading Mackay's book, one is left with the firm impression that human beings were, at that time, an unbelievably stupid and avaricious lot. The Canadian mining bubble proves that nothing much has changed. It carried on for years, culminating with Bre-X, the promotion that fooled some of the most powerful, and ruthless, people on earth. The world's largest mining companies and their all-star boards of directors became consumed with their deluded, underhanded attempts to grab control of Busang, Bre-X's fabled gold deposit in Indonesian Borneo. When members of Indonesia's corrupt ruling family began squabbling over Busang, it threatened to unleash a political firestorm, during a crucial period of presidential succession. Fortunes were risked; lives were left hanging in the balance. *Yet no one thought to question whether the gold actually existed.*

A confession: I believed. For too long, I thought Busang was real, not simply because it was so huge, and important, and because the biggest names on Bay Street spoke so convincingly about it. I believed in Busang because it was a good story. I'd heard the mining industry was full of boozy cheats and unscrupulous promoters, although I didn't realize how far they reached into "respectable" society. I soon discovered that the exploration business bears more resemblance to professional boxing than anything else.

On my first visit to Indonesia, in December 1996, Busang was being touted by the investment professionals as the world's richest gold discovery. Initially, my attention was fixed on the unseemly corporate and political battle over this allegedly fabulous deposit. The story had changed completely when I returned to the archipelago four months later. Busang had just been revealed as a hoax. The implications suddenly grew exponentially, both in Indonesia and in Canada. As things began to unravel, and nerves were exposed, and people vented their anger and their grief and their fear, I found myself walking inside a much bigger drama. Bre-X was only the tip of the iceberg.

The lies didn't begin with Bre-X, and they didn't die with it, either. Junior exploration companies routinely use false and misleading statements to encourage investment. They tweak drilling results, manipulate data to make their properties appear more favourable, and fail to disclose material information. Some resort to sprinkling gold dust in their drilling samples. They do it because it's easy. Promoters, geologists and investment touts know the score; deposits are salted to make money. If they get caught, they get their knuckles rapped and move on.

The agencies that regulate Canada's stock markets are reluctant to enforce their own rules. They act on behalf of their members instead of the investing public. Promoters found guilty of trading infractions and other indiscretions are given all kinds of breaks by the regulators and are seldom punished. Sometimes, they end up in charge of other companies.

Mining companies and the investment analysts who cover them are involved in symbiotic relationships that favour each other, and not investors. The investment pros rarely undertake any thorough technical evaluations or corporate due diligence. Brokerages that advance a mining company's stock get preferential treatment and financial rewards.

Instead of acting as watchdogs, media regurgitate what mining companies and analysts tell them and rarely investigate even the most outrageous claims. Journalists who do expose corruption are dismissed as irresponsible muckrakers, and are even threatened with violence. And so the worst companies carry on, conducting themselves poorly outside of the country, and are rarely taken to task. Some of Canada's major mining companies use bribes to curry favour with foreign governments. Others are involved with environmental degradation and the abuse of human rights — all in pursuit of cash.

Incredibly, there's been no word of an inquiry following the latest series of frauds in the junior mining sector, and no loud calls for reform. What's outrageous about the whole fiasco isn't just that we allowed it to happen in the first place. The real scandal is that it is bound to happen again. But what is past is prologue, something Charles Mackay knew all too well. His own work, he noted, "may be considered more of a miscellany of delusions than a history — a chapter only in the great and awful book of human folly which yet remains to be written." This is the latest instalment.

REQUIEM

*The government this year resorted more and more
to the 1963 anti-subversion law, a powerful and all-
encompassing legislation that has proven effective in
clamping down on critics. The law carries a maximum
death penalty.*
– *Jakarta Post*, December 1996

*It would be better for us to just forget about Busang.
It was a nightmare.*
– Mohammad Sadli, Indonesia's former minister of Mines
   and Energy, May 1997

FRIDAY, 30 MAY, 1997. JAKARTA, the day after
Indonesia's violent national election. The largest
city in Asia is back to its usual simmering calm
after three weeks of riots, stabbings and fires. More than 260 people died during the latest campaign, but for Golkar, Indonesia's ruling "functional group," the victory is sweet. Scripted to win, Golkar
claimed seventy-four per cent of the votes, even better than targeted.
It was a message from President Suharto: His authoritarian grip on
the nation remained firm. "Only with divine intervention can

Golkar's support slump," one national scholar noted before the vote. Suharto hadn't taken any chances. Indonesia's official opposition parties were seeded with troublemakers. Four million civil servants were instructed to vote Golkar or lose their jobs. People pressing for change were cornered and jailed. The election was over before it had started.

Down at the Smuggler's Arms, politics is the last thing on anybody's mind. A party is about to start. Two hundred expatriate engineers, geologists, rig-pigs, mappers, office managers and assorted guests are gathering in this private poolside bar for the mining community's monthly get-together. It's the eighteenth "evensong" and beer bash, organized by the Society of St. Barbara, the patron saint of miners and firemen. Three years ago, St. Barbara blessed her flock with a miracle: a massive gold discovery in a remote Indonesian province on Borneo. It was called Busang. This miracle grew bigger and bigger, attracting geologists, money men and investors to Indonesia like ants to honey. Dozens of opportunistic companies parachuted into the country and hammered stakes around Busang. Eight billion dollars was raised for exploration, making St. Barbara's worshippers very grateful indeed.

Tonight, they come not to praise Busang, but to try to bury it.

EXCLUSIVE INVITATION: Our forthcoming *Evensong* will be a requiem. Like all requiems it is a sad affair for our mining community, **Dearly Beloved**, and a time for introspection. The Busang gang broke almost every one of the Ten Commandments, especially the first: 'Ye cannot serve God and mammon' . . . .

It was a cynical joke, a way to release the tension. The invitations were faxed to all the usual suspects. Someone has dropped a large bag of salt at the bar's entrance, a poke at the tampering done at Busang. Worthless scraps of core, once thought to contain a rich grade of gold, are handed out as door prizes. There is great laughter as Tim Scott, a burly administrative officer with Barrick Gold Corp., recites a limerick denigrating Bre-X Minerals, the tiny Canadian exploration company that touted Busang as a geological wonder.

There are a lot of world-weary Australians here, like Scott, and they enjoy his sarcastic wit. A few Indonesians mill quietly in the crowd. Naturally, no one from Bre-X shows up. The people who milked the biggest scam in mining history are long gone. The grunts have all scattered. Management has split. David Walsh, Bre-X's flustered promoter, is claiming outrage and building a legal defence back in Canada. John Felderhof, his chief geologist and former pal, opted for safe harbour in a Grand Cayman fortress. Michael de Guzman, Bre-X's exploration manager, is presumed dead after allegedly throwing himself from a helicopter, panic having suddenly seized him after three years of rampant and artful deceit.

The miners left behind pretend everything is fine. Bre-X was a passing squall, they say, a freak. But no one believes it. Bre-X created a mirage and it is evaporating around them. The Department of Mines is still torn by competing interests. Bribes move back and forth; no one is sure who controls what anymore. Most of the thirty Canadian exploration companies that rushed into Indonesia behind Bre-X are making plans to pull out. The good times are over.

The kegs empty, the noise level rises, nerves are exposed. A few arguments break out. There's some pushing and shoving, nothing serious, but enough to make Sister Endang, the bar manager, shut off the free taps. The bar is swarmed; soon it's impossible to shove through the crowd to buy a beer.

I bump into an American lawyer, Tom Ajamie. He's in Jakarta for the third time in six months, doing the groundwork on a class-action suit against Bre-X. Ajamie, who bears a striking resemblance to the actor Kevin Spacey, is young, laid-back and smart. Based in Houston, he's worked on mining cases before; in 1996, he helped settle a major ownership dispute over the Voisey's Bay nickel deposit in Labrador. Later, he represented a Houston high roller who was burned investing in a bogus mining outfit. The company had allegedly paid stockbrokers to promote its phoney claims. Ajamie managed to get a surprise confession from one broker, who burst into tears and admitted to having touted the company in exchange for a $15,000 "loan."

Ajamie's experiences taught him that the mining scene is a predatory place, where the space between truth and fiction is often blurred. But he was unprepared for Bre-X. It involved an unprecedented deception, a $6-billion fraud. Someone supposedly tricked the sharpest investors in London, Toronto and New York into believing that Busang contained more gold than any other piece of real estate on the planet. When the hoax was exposed, the company's top brass denied any involvement and pointed the finger at de Guzman.

Ajamie wasn't buying it, nor were hundreds of small Bre-X investors. They sent him to Indonesia to play detective, and while he doesn't relish the role, he's good at it. We'd never met, but he picked me out of the crowd at the Smuggler's Arms. He suggests we cut out. A couple of Aussie drillers are willing to talk about Bre-X. We head out the door and go into the night. It's warm, damp and dark.

Ajamie drives off with a personnel manager from one of the local drilling companies, while I follow behind in a private taxi. My driver speeds through the narrow streets, trying to keep up. Eventually, we stop at a large gated house. A couple of men squatting in the driveway point the way inside and it's suddenly very bright, all white marble and stainless steel, high ceilings, with a large swimming pool in the back. The two rig-pigs, Peter and Wayne, are already hunched over a large table in the dining area, devouring a plateful of chicken. Back in Oz, they'd probably just be scraping by. Here, they get a taste of the good life. Their families are happy in their glitter domes and are not keen to leave. "The guys at Bre-X got rich and then they ruined everything for the rest of us," Peter says, cracking open a beer. "No one is going to want to sink any more money into this place. What do you think that means, eh? Bre-X is going to put us out."

Wayne is drunk by the time we get to the Scotch. He's banging on the table and yelling at us. "It's the fucking North Americans who ran this scam," he screams. "You think we all knew what was going on? What about you?" Wayne is a large man, and his tone is menacing. It's three in the morning. My elbows are sliding off the table. It's time to go.

My driver takes us north, towards downtown. I ask Ajamie if he enjoys working on the Bre-X file. He just laughs. "People hear that I'm working on Bre-X and they're like, 'Hey, you got yourself a slam dunk,'" he says. "It isn't like that at all." Ajamie has to turn up some solid evidence that Bre-X, its technical and financial advisors, and three of the world's largest brokerages knew — or at least should have known — that Busang was a scam.

"This is the hardest case I've ever been involved with," he tells me. "Frauds can be tough to prove. Then you add the Indonesian component. How do you know who's lying, who's got an agenda? This place is all snakes and ladders."

He gets out at the Four Seasons and the driver veers into Menteng, an old, central district where most of Indonesia's political and military elite live. Ten blocks from President Suharto's house, we pass a gaggle of transvestites. "Lady boys," my driver giggles. During the day, the street is lined with men selling dogs, mangy little creatures locked inside cages.

Jakarta is not one of Asia's gems. It's a battleground. A grimy port city on the island of Java, founded fifteen centuries ago by Hindus, it was stormed by Muslims and later razed and remodelled by Dutch traders. The Japanese army invaded in 1942 and ran amok for three years. After the war, the Dutch reasserted their power until nationalist forces seized control and went about building the city into the heart of a new, militarized nation, strewn five thousand kilometres across a bountiful, ethnically diverse archipelago. Comprising more than thirteen thousand islands, Indonesia is home to 200 million people representing myriad linguistic and cultural traditions. Half the nation's population lives on Java. Rapid, unfettered growth has seen Jakarta's population rise from 500,000 to 10 million in just fifty years. Infrastructure is wanting; the roadways are jammed, the sidewalks are crumbling, and it's just getting worse. Moving around in the heat and humidity and pollution is a chore. When I try to relax with a row around the stinking harbour, I end up with a painful ear infection. Tim Scott can't hide his amusement. "You're bloody lucky you didn't catch bronchial syphilis," he roars.

Disparities bombard the eye. Gorgeous, shimmering office towers line its main streets. Soldiers armed with machine-guns stand guard outside the walled gardens of the rich and powerful. Ragged children listlessly flog sticks of gum in the streets and wash themselves in fetid canals, sorry vestiges of the city's colonial past. These open sewers also serve as laundromats, garbage dumps and a source of drinking water for Jakarta's most wretched citizens. Surrounded by squalor, they are encouraged to accept their lot without protest.

Raising one's voice only brings trouble. Dissenters are dealt with harshly. Jakarta is a bed of paranoia. Power and wealth are doled out to Suharto's cronies and stripped from the unsuspecting. "This is a president who has been in power for thirty years. Who's not afraid of that?" explains a career civil servant, a Western-educated "technocrat" who is a favourite of Suharto. "Even me, I am afraid," he says. "We have to always be very careful. This is still a military country and anything can happen."

The technocrat has been an excellent source, with first-hand knowledge about the complex political struggle that upstaged the Busang sleight of hand. An interview he gave me months earlier angered the minister of Mines and helped cost him his high-level job inside the department. He shrugs it off. What's really important, he says, is that he maintains his good standing with Suharto. But I can see that his manner has changed. He seems edgy, and he's constantly clearing his throat. But his story is still the same. Fraud or no fraud, Busang was used by some of Jakarta's elite as a means to enrich themselves and their friends.

That evening, he calls me at my hotel. "Look," he whispers, "I made two mistakes today. The first was seeing you in person. The second was letting you tape our discussion. People may be watching me. My life could be made miserable if anyone heard our conversation. Promise me you'll destroy the tape." I tell him I'll look after it. "No, don't look after it, just destroy it. And listen, you had better look for a new hotel." He hangs up before I can ask him why.

Post-Busang, we have the dead — Mike de Guzman, the half-dead — those who will be living in a legal or financial purgatory for years to come, and the living — like most of us, *Dearly Beloved*, also exisiting in some sort of demi-world, unsure of what the future holds . . .

"You want to go to Busang?" Simon Sembering takes a deep drag on his cigarette and shakes his head. "I would advise against it. This is not a good time." It's no surprise. Sembering, a middle-ranking bureaucrat inside the Department of Mines and Energy, would rather everyone forget Busang. It had symbolized a rich and prosperous future for Indonesia. Now it only stirs resentment. During the election, the Bre-X scandal was a sensitive issue. Prominent Muslim nationalists suggested that the government had sold out to greedy foreign hucksters. They said that few Indonesians ever benefited from the exploitation of natural resources, which were supposed to belong to everyone. Students demonstrated outside the mining minister's office downtown, drawing attention to the scandal until they were removed by the police.

Fifteen hundred kilometres to the east on the island of Borneo, in the Indonesian province of East Kalimantan, hundreds of migrant workers were drifting down the rivers from Busang, dejected, their pockets empty. An angry group of men stayed behind, demanding severance packages. Busang, once Bre-X's private domain, was now under the authority of the Indonesian government. Soldiers had secured the area, keeping it free from foreign journalists and other unwelcome intruders. "You would be taking a chance to go there now," Sembering tells me. I thank him and leave. Spending another week in jittery Jakarta was hardly appealing. Three days later, I land on Borneo.

Samarinda is the closest major city to Busang. It's a calm, easy place, tucked a few kilometres inland from the Makassar Strait. Most foreigners come here to cruise the Mahakam River, Indonesia's Amazon. A broad, winding waterway, the Mahakam has served as a

vital commercial route for centuries, spiralling one thousand kilometres into the rainforest. Now that a fledgling tourist trade has
slowly developed along the muddied river and its tributaries,
Samarinda is awash with determined hawkers pitching houseboat
cruises. The best river craft are stocked with wet bars, hot showers,
air-conditioning, all the creature comforts one would expect to find
in any big-city hotel. Travellers can spend weeks floating up and
down the rivers of East Kalimantan, gazing at the passing jungle
scenery, eating fresh prawns the size of a fist. Local guides are quick
to recommend the longest, most luxurious tours.

Young, clean-cut, dressed in crisp cotton shirts with button-down
collars and neatly pressed chinos, Dian and Kani look like a couple of
computer programmers, not the "jungle adventure guides" they bill
themselves as. But they know the territory and its people and can
guarantee me passage to Busang. The site is 150 kilometres to the
northwest, as the crow flies, double the distance on land. A helicopter
would take just thirty minutes; a speedboat, perhaps four hours. But
the cost is prohibitive. There is a third alternative, Dian finally allows,
much more "traditional," and far cheaper. It might seem a little basic,
he warns. Do I mind walking? Sleeping on the ground? Can I operate
a motorcycle? What is my honest opinion of snakes?

Kani quietly traces a route on a map with his finger. I ask him to
point to Busang. He hesitates. It is obvious he has only a rough idea.
"Don't worry," he laughs. "We can always ask for directions. We'll
get you there in a couple of days." He peers outside. "Just pray that it
doesn't rain."

It pours most of the night. At eight the next morning, Dian and
Hani show up inside a battered four-wheel drive. The vehicle's
owner, a tall, skinny chain-smoker, professes an intimate knowledge
of the logging roads we need to follow to get to the Kedang Kelapa
River, a tributary of the Mahakam. From there, we will hire a craft to
Long Tesak, the closest village to Busang accessible by water. We'll
sleep in Long Tesak and continue over land the next day. Our final
destination lies somewhere beyond in the jungle.

Soon we're out of the city and into the hills, past the bright green

rice fields, then fishtailing down a private logging road, gears jam-
ming, red mud flying over the hood of the truck. The air-conditioner
is broken and Dian looks like he's going to be sick. I begin to wonder if this is a mistake. My two guides haven't bothered to bring
any food or camping gear along. We would find food and shelter
along the way, in the native communities, explained Kani. There are
roughly two hundred Dayak tribes in Borneo, all with their own traditions and dialects. The Kenyah, who dominate the region around
Busang, are considered among the island's most culturally advanced,
with an intriguing reputation as headhunters. "Don't worry, they
gave it up a long time ago," Dian jokes. "Otherwise we'd all be in
trouble. I'm Tunjung."

Dian grew up in a tiny village on the Upper Mahakam, called
Depecoping. His ancestors used to war with the Kenyah. These
days, his people are struggling to survive the effects of moderniza-
tion. People are leaving for the cities. Dian, twenty-nine, has firmly
rejected the old ways of subsistence farming and is raising a family in
Samarinda. He is the first from Depecoping to have attended university, and like Kani, he speaks excellent English, along with Bahasa
Indonesian, the national language, and several local dialects. His
parents, who grow rattan on the banks of the Upper Mahakam, have
always been poor. Dian has other ambitions. "When I was young, I
hated working in rattan," he says. "I didn't want to stay in my village
for the rest of my life. I want to buy my own business." Dian and
Kani represent a new, growing segment in Indonesian society, the
secular middle class. Unlike their parents, their values are distinctly
Western. Like a lot of young Dayaks, Dian sees the development of
Borneo as a plus. He's heard all about the Bre-X scam. Michael de
Guzman's death seems to fascinate him. He's anxious to go up there
and see what's left.

First we must run the gauntlet. Our initial obstacle is a military
checkpoint at the edge of a private timber concession. The army is
practically everywhere in Indonesian Borneo, trying to keeping illegal loggers at bay. Half of East Kalimantan's seventeen million
hectares of rainforest have been targeted for harvest, a situation that

has meant hardship for entire Dayak communities. Some have been physically removed from their plantations in order to make way for new timber concessions. Compensation is pitiful, in many cases non-existent. Villagers have reportedly been forced into labour contracts with the logging companies. Others have resorted to "squatting" in the forest and poaching trees. The government responded by banning the sale of chainsaws. Now the army mounts regular patrols through the province, ferreting out trespassers.

Our driver, Dany, negotiates briefly with some soldiers standing outside a hut. It will cost us 40,000 rupiahs, about $20, to pass. The road narrows as we drive farther into the forest. The route is flooded; Dany has to struggle to keep the truck on course. We skid around a bend and stop. A pond ten metres long stretches in front of us. Another truck sits there, idling, its driver evaluating his chances. We egg him on. Kani and some men from the other truck wade into the swamp. The water goes up past their belly buttons. They poke around, looking for the shallowest parts of the miniature lake, and flag a crooked path with sticks. The driver guns his engine and his vehicle lurches into the water, getting halfway across the divide before sinking deep into the mud.

With the road now blocked, we've got no choice but to walk. First, however, we strip down to our shorts, slide into the water and start pushing the trapped vehicle back to shore. It isn't easy; the mud underfoot is slippery. It's like running on a bar of soap. Kani walks off down the road, and comes back with a scrap piece of wire — a hauling device. An hour later, we finally free the truck from the pond and set off on foot, lugging our backpacks. There are nine of us now, including the passengers from the other truck. We must look like a bedraggled band of refugees, squishing through the chattery jungle in our underwear, mud caked to our legs, our hair, mud everywhere. It's not long before a small pick-up appears in the distance; we rush to it, throw some money at the driver, and pile in the back. The truck reverses direction, and we are off once again, the sun beating down on our backs. My skin is burning. We have just crossed the equator.

Mercifully, we reach the first river in less than an hour, and soon we are skimming comfortably along the Kedang Kelapa, a meandering, narrow river. We've hired a long, narrow craft called a river canoe, with skipper and 20-horsepower engine. I find some shade under the canopy and settle back. The water is smooth as glass. The breeze feels good; I wave my toes at the shoreline and start to relax. We pass some small wooden houses on stilts. Long, chinked logs serve as stairs down the river's steep banks to floating outhouses. Some have little concession stands. Women and children squat on the floats, scrubbing clothes. Their faces are smeared with a rice paste, protection from the sun.

We encounter other passenger boats travelling in the opposite direction. Barges slowly drift by, carrying cargos of rice. Everyone stares as we glide by; it must be rare to see a Westerner in a river canoe. At Maura Angcalong township, we swing onto a bigger river, maybe seventy-five metres wide, and choppy. Kani taps my shoulder and points to some enormous beehives hanging from the trees. We float past fruit plantations and hibiscus, water buffalo and great horn-billed birds, surreal-looking creatures that the Dayak consider magic. Although most of the Dayak have converted to Christianity, they have tried to retain some of their animist beliefs.

Before dusk, we land at Long Tesak, a humble collection of weather-beaten wooden huts. A handful of villagers stands on the river bank, pointing and giggling as we cling to their toilet float. Kani scrambles up the slippery gangway and makes some inquiries. "It's important to find the headman and see if we can spend the night with him," Dian tells me. We are in luck. Kani reports that we've been invited to stay. We grab our stuff and walk down a long grassy path to Ngang Bilung's house.

Long Tesak does not appear prosperous. The fifty-odd houses are worn and primitive. The headman's home is dark and shabby. A few wobbly pieces of furniture huddle against a wall in the main room. The walls are decorated with calendars bearing the golden Bre-X crest. A picture of Christ hangs over a battered desk. We go through a small courtyard and into a kitchen area that's cluttered with sacks of rice, piles of dried beans, motor parts and fishing nets. There's an open fire used for cooking. Since Ngang Bilung is headman, his home also boasts a gasoline-fired generator and a cement toilet, a dicey-looking contraption but handier than the alternative down on the river.

Dian and Kani go off to collect some food for dinner, leaving me alone with the headman. He offers me a cigarette. His hands are rough, the nails thick and chipped. I look at him closely. He's a small, wiry man, probably in his fifties. His undershirt is filthy. His teeth are stained with tobacco juice. I must look ridiculous. We eye each other and smile.

Ngang Bilung's wife returns from the rice field and makes welcoming gestures at me. Their three teenaged sons move about quietly, grinning shyly. Dian and Kani come back with some eggs, green beans and tins of sardines. They crouch on the floor, frying things up with some rice. We eat with our hands. Ngang Bilung gently asks what has brought us to Long Tesak. Dian tells him we want to see Busang and asks if he can help us find our way there. Ngang Bilung says he thinks he can.

After dinner, Ngang Bilung fires up the generator. A couple of light bulbs dangling overhead begin to glow. After seating us around his battered desk, he produces a plastic water bottle full of cloudy rice wine. It is sour, slightly fizzy, and packs a decent punch. We present our gifts, some black, sticky plugs of chewing tobacco and a bottle of wild honey.

A handful of village elders drops by, and they take their spots around the table. One of the men has huge, elongated earlobes, with holes in the middle, big enough to accommodate a golf ball. I ask him what they are for. "Without my long ears, I would not be acknowledged by my family," he says, Dian and Kani double-teaming with the translation. "They might think I was a monkey." Protestant missionaries who came to Long Tesak at the turn of the century discouraged the Kenyah Dayak from decorating their bodies. Although a few of the older villagers sport elaborate tattoos and the long, distended ear lobes, younger generations have given up the practice, not out of disdain, Ngang Bilung points out, but because "they are afraid that kids from the city will laugh at them."

The city creeps closer. The elders say they are worried about their animistic, agrarian culture, seventy per cent of which has been lost, by their estimate. There is little room for subsistence farming in Suharto's New Order regime. Jakarta views the Dayak's land as a prime commodity, and their villages and rice fields as obstructions to development. A resettlement policy, which has seen thousands of Indonesian labourers from across the archipelago moved to Borneo's three Indonesian provinces, has caused serious cultural tension. The problems that have caused unthinkable violence in East Timor are beginning to take root here.

The elders ask Dian and Kani about the clashes on the west coast, where Muslim settlers from the island of Madura and the indigenous Dayak have been battling over territory. Fighting broke out among the two groups at the end of 1996; the Indonesian press estimated twelve hundred people were either dead or missing after a month of violence. Dozens of gutted and headless Madurese corpses were hung from trees outside government offices. The army tried to put a lid on the disturbances, sending three thousand troops to West Kalimantan, but that hardly helped. In February, soldiers gunned down two dozen Dayak attempting to push their way through a roadblock. One soldier was hacked to pieces by hatchet-wielding natives. Ngang Bilung and the other elders shake their heads. Their world is falling apart.

Busang represented another threat, not a boon, the elders say. Three large-scale mines operating within 250 kilometres of Long Tesak have had a devastating impact on local people. Effluent has polluted their rivers and destroyed aquatic life. Forests have been cut back for open pits, roads and airstrips. Crime and prostitution have been introduced. The Kelian gold mine, opened in 1992 by Australian conglomerate CRA Ltd., expropriated Dayak land, bulldozed two villages, and ousted four thousand local miners who had been panning gold in the area for decades. A year later, six hundred Dayak demonstrated against the expropriation of their land. A dozen men were arrested by the military police; one died in custody. According to Indonesian environmentalists and human rights advocates, Dayak from seventeen villages downstream from the Kelian mine "often suffer rashes, eye infections, and nausea. This is caused by leakages from the [mine's] waste pond, which contains cyanide."

In January 1997, another Australian company encountered problems with "illegal miners," its term for local Dayak men and women who have been stripped of their long-standing right to mine gold. According to Pelsart Resources NL, which owns the Ampalit gold mine in Central Kalimantan, "Military personnel were mobilised to seize illegal equipment and move several hundred miners from the site. While this move was generally successful, it met with some resistance in which several trucks were damaged and a field office building was destroyed in a fire which was deliberately lit. While some illegal miners have returned to the site, the numbers are currently being kept under control by the presence of mine security assisted by the military presence in the area."

Big gold mines present big problems; the people of Long Tesak know this. "We were not happy when Bre-X arrived," Ngang Bilung says. "Busang is only thirty kilometres from here. The men used our river access for their boats and barges. If they built a mine, we knew we would lose some of our land, our rice fields and our gardens. The river would be dirtied. I asked Jerry Alo [Busang's site manager, a Filipino] for some support. He said that Bre-X is not responsible for our village. So I closed the village to them. Closed it off. After a while, they came back and said, 'Let's talk.' They promised us jobs and education and said they would supply us with clean water after building their gold mine. So we felt happier and we decided to welcome them. And some of us did get jobs, hard jobs, digging with shovels. Others sold the company vegetables and chickens. But it wasn't what was promised. And suddenly everything stopped."

I ask Ngang Bilung what he thinks went wrong. "The men from Bre-X did not act properly," he says. "You know, there is gold here. We have been mining it our own traditional way for decades. But

Bre-X did not give thanks or ask the spirits for blessing and support. They should have done a dance ceremony, offered some animals. That is why they did not find any gold. They showed no respect for sacred things."

> They forgot that the Almighty is a jealous and vengeful God. No wonder, **Dearly Beloved**, we are seeing an exodus . . .

We sleep on the floor that night, next to a burning mosquito coil. Dogs and chickens rustle around underneath the floorboards. In the morning, Ngang Bilung bangs out a letter on his ancient typewriter, our ticket past the guards who man a checkpoint on the road to Busang. He also instructs one of his sons to go along with us. The four of us set off down a muddy path that runs parallel to the river. The route is lined with banana trees, coffee plants and pineapple plants. We soon come to a river landing, where a pair of Bre-X speedboats are tied to a dock. Kani hands our letter to a security officer, who looks at the paper impassively. After some discussion, he motions to a white van. One more hurdle is overcome; we are almost there.

Our new driver's name is Beni. He shoves the van into gear and we are off on another heart-stopping run, barrelling around blind corners and down steep straightaways, barely slowing down as we meet oncoming trucks overflowing with workers leaving Busang, this time for good. Three-quarters of an hour later we lurch to a stop at the base of a hill. Spread out beneath us sits a strange sort of township. Row after row of small, wooden houses are laid out in a large clearing. The buildings, more than eighty in all, are identical, about three hundred square feet in size, painted a dull beige. They are squat, sterile and utterly foreign to the natural jungle environment.

I'm struck by the eerie stillness of the place. Mosquitoes buzz around our ears; that is the only sound. We're wrapped in a blanket of heat and humidity, not a whiff of breeze. A few faded pieces of laundry hang limply from a clothes line, the one lingering sign of human activity. The scene looks like an empty film set, built to impart a false sense of scale and permanence. Which is not far from the truth. A deserted company town, built to accommodate hordes of Indonesian workers: this is what's left of Busang.

Kani spots someone darting into one of the little houses, and we decide to investigate, going down the hill and past a small white mosque. Inside one of the huts, a rail-thin, thirty-year-old man is sitting on a plastic-covered floor. He is agitated, pulling at his fingers. His name is Hidiar. He and his wife were among the first Indonesians to migrate to Busang. In February 1995, they came from Sulawesi, an island east of Borneo, after reading an advertisement Bre-X placed in their local newspaper. "It said the chances for long-term work were good," Hidiar says. "We sold everything and left our families. When we arrived here, they put us in a tent. We lived in it for more than a year, and then they built the houses. We had electricity. Flush toilets outside. This was the best place we ever lived."

Almost a month has passed since Busang was declared a fraud; Hidiar is still here because he's got no place to go. He drove a tractor for Bre-X. His pay was 400,000 rupiah a month, about Cdn$235, twice the national average. But the working conditions were brutal. He routinely logged twenty-hour days, building roads and clearing swathes of rainforest for the Bre-X drill crews. He did not receive any overtime pay. Hidiar and his fellow Indonesian labourers despised most of the Filipino geologists who managed the Busang project, especially Jerry Alo, the site manager. "He treated us badly," he says. "Like we were less than him." When the work stopped, Hidiar's friends seized two of Alo's underlings and locked them up in a room. "We wanted them to feel like us, with no respect," Hidiar says, with a grim smile.

Now the power has been shut off and the outdoor toilets are overflowing. A clunky stereo system Hidiar bought from another worker is stacked in a corner, useless. I ask him what he plans to do next. "I have no idea," he says, staring at his feet. "The dream is gone." It's the same situation for hundreds more Indonesians who migrated to Kalimantan, seeking work with other Canadian exploration companies. Once Bre-X collapsed, most of the junior mining outfits reduced their activities in the area or abandoned the area completely.

We spend the rest of the day roaming around the area. The main mining camp lies a couple of kilometres down the road. Two stone-faced security guards carrying odd, curving knives hanging from their belts refuse to let us past the gate. They won't be bribed, so we continue a little farther along the road until we come across the local "prostitution complex," a solid new building made of bamboo, palm leaves and roughly hewn timber. It features a veranda bar, a long hallway and six little rooms. The madam serves us warm beer and rouses two sleepy-headed prostitutes who have been napping in the midday heat. They sit at the table, smoking and scratching themselves as the madam tells her story. Business, she complains, has been terrible. The complex is not even two months old, she says. It opened just before everyone packed up and left. She will never recoup her investment. "Maybe I'll go and start a rice farm," she says, throwing up her arms. The two prostitutes scream with laughter.

The chief of security arrives on his motorcycle, smiling, waving at the girls and slapping one on the rear. Erman Bratha has heard there's a journalist looking around; he has come to make me an offer. He will take my camera into the work camp and take pictures, in exchange for a "small gift." I thank him and tell him I'd hoped to go inside myself. He looks at me dubiously and orders a Guinness.

Despite his casual manner, Bratha is the consummate company man. His loyalty to Bre-X is unwavering; he insists that there is gold at Busang, that certain Indonesian interests have finally grabbed the property from Bre-X. "David Walsh, he is a good man," he tells me. "David Walsh would do nothing wrong." I ask him if he has ever met the portly Bre-X founder. "No, but I saw him once. He was here, ha, he has a big belly. He made us jobs."

Bratha says he will stay at Busang until he is ordered to leave. The few remaining Bre-X employees, he says, "are all like chicks without a mother hen." There is no sense of order any more. The camp is overrun with forensic accountants and private investigators. "Everything is so strange now," Bratha says, looking at me from across the table. "I wonder, in Canada, is it like this?"

2

THE ANALYST

The good times won't last forever. But when you've just pocketed a $1 million bonus, who cares?
— *The Globe and Mail,* 12 December, 1997

[Investors should] worry if all the analysts following a company rate it a "strong buy."... These guys are like everybody else: they feel more comfortable as part of a majority. It's the old story: if you are wrong and everybody else is, too, you get little blame. Nobody loves a party pooper.
— *Forbes* magazine, 15 December, 1997

THEY APPEAR en masse every lunch hour, smartly dressed men and women with cellphones stuck to their ears, spilling into the granite plazas that line Bay Street, the epicentre of Canadian finance. These are the Modern MBAs, bright young brokers, analysts and accountants, in fast pursuit of cash. The market is humming; money has been pouring into the Street ever since the last stock market run ended abruptly back in

1987. The Modern MBAs, their eyes trained on the bottom line, have seen investment in the Canadian stock market grow twelve-fold since the crash of Black October. Many missed that crash; they have never experienced a serious downturn in the economy.

The competition is intense, thrilling, addictive. The Modern MBAs scour the country, looking for companies with sexy "stories" they can sell to the public. A positive research report from an analyst might land his firm a lucrative assignment, underwriting a new stock offering. Brokerages and investment houses reap big bucks dealing a company's fresh stock to investors. Initial public offerings (IPOs) raised $14 billion in 1997, almost triple the amount raised in 1995, producing billions of dollars in commissions. Mining accounted for more IPOs than any other sector in 1996 and 1997. Whether investors make money is secondary; the brokerages, as middlemen, always win. And so it goes, find a story, flog a story, buy and sell. As long as there's faith in the marketplace, the modern MBAs are secure, strapped on top of the rampaging bull.

Bill Stanley used to feel the same way, until the Bre-X juggernaut swept him off his feet and took him for a long, twisting ride through the dark side of corporate Canada, finally dumping him off on the sidewalk, battered and bruised and more cynical than ever. Stanley cleared over $1 million from Bre-X, more than enough to build a dream house for his family, deep in the pastoral solitude of suburban Mississauga. It weighs heavily on his conscience.

Stanley is haunted by Bre-X. Months after his Bay Street brethren have moved on to the Next Big Story, he sits in the commuter train every morning, playing the fraud over in his head, starting with the first meeting with Bre-X in Toronto, to that weird, unsettling trip he made to Busang with the other Bay Street Pros. Looking back, he can see how John Felderhof and Michael de Guzman showed them a good time down in the jungle, laughing and drinking, making them feel as though they were privileged and important and witnessing something historic.

But there were those troublesome tics — worried voices, nervous looks, unanswered questions — all of which suggested something wasn't quite right. Eventually, Stanley began to wonder out loud. His bosses at the Bay Street investment bank where he works told him to get over it or move on. Stanley bit his tongue, but he never shook the bug.

I meet Stanley in a crowded bar underneath First Canadian Place, the tallest office tower in Canada, and headquarters of Nesbitt Burns Inc., the most enthusiastic and favoured underwriter of Bre-X stock. Stanley's choice of venue was inspired. The bar is probably swarming with Nesbitt types who've been, as he put it, "busanged."

Stanley is in his mid-thirties, tall, with steely blue eyes. Although he wears standard Bay Street issue — shiny black shoes, dark suit, monogrammed shirt, little golden bull cuff links — his loopy grin suggests a certain unorthodoxy. "Welcome to the den of iniquity," he smirks, lining up a pint of draft before launching into his story. There's just one condition. "You can't use my real name," he tells me. "If you do, I'll get fired, and then I'll have to kill you." He sits back and gives me a wry grin. "Ever flown over the jungle in a helicopter?"

Like just about every young suit on Bay Street, Stanley sports an MBA. He graduated from business school in 1991 and landed at a mid-sized investment house, working on mergers and acquisitions. A couple of years later, he joined another firm, this time as a "special situations analyst," responsible for finding undervalued assets worthy of investment. Despite his fancy job description, Stanley's salary was surprisingly modest. He drove a rusted pick-up truck and lived with his wife and two small children in a cramped condo downtown.

He first heard about Bre-X in 1995. No one knew much about the company, or what it was doing in Indonesia, but its stock was hot, jumping from $2 at the start of the year to $15 by the end of the summer. In September, Stanley went to a Bre-X presentation at the Royal York Hotel. David Walsh, the president, was there, pressing the flesh with his exploration manager, John Felderhof. Stanley had heard a few things about Felderhof, that he had discovered a huge gold deposit way back in the late 1960s, that he knew

Indonesia better than anyone. Now he was in Toronto, talking about volcanic centres and diatreme dome complexes. "Most of it flew right over my head," Stanley says. "I'm not a gold expert or a geologist. But Felderhof was talking about a ten-million-ounce deposit. And I knew that was a lot of gold."

Bre-X was Canada's third spectacular — and unlikely — mining story in four years. In 1991, a small company found diamonds in the Northwest Territories, setting off a major exploration rush. A Canadian geologist named Charles Fipke, who'd spent a solitary decade prospecting north of Yellowknife, announced he'd found eighty-one tiny diamonds in some kimberlite rock near Lac de Gras, a barren, windswept piece of tundra. Five months later, Fipke uncovered more diamonds, bigger and more impressive than the last bunch. Some stock promoters hailed it as a $100-billion find, and Fipke's company, Dia Met Minerals, saw its shares balloon from a low of 35 cents to $67. Prospectors from every corner of the planet turned up in Yellowknife, creating a staking frenzy not seen since the days of the Klondike. The world's largest diamond producers arrived on the scene, snapping up property and signing joint-venture deals with local exploration outfits. Two large diamond mines are scheduled to begin production by the new millennium.

Three years after Fipke's find, a diamond hunt in Labrador led geologists to a vast nickel deposit. The flukey discovery by Diamond Fields Resources Inc. made a legend of Robert Friedland, a gangly flower child turned stock promoter, when he sold the property to Canadian nickel mining giant Inco Ltd. for $4.3 billion. Friedland was hailed as a mining genius, says Stanley, "even though his previous record was shit. If anyone had bothered to look, they'd have seen that David Walsh's reputation was shit too. But Diamond Fields was on everybody's mind. Nobody wanted to miss out on the next one."

Bay Street turned Bre-X into a major company. Large public institutions such as the Ontario Teachers Pension Fund and the Alberta Treasury were buying stock. Mutual funds were gobbling it up. Soon the American traders would be all over it. There was lots of big-ticket underwriting to be done, plenty of buying and selling, loads of profit potential. The last thing anyone wanted was some analyst to ruin the party by dumping on Bre-X. "Imagine you write a negative report," Stanley says. "David Walsh gets on the phone to your boss and screams at him. Everyone else is positive, he says. He'll never do any business with the firm. The analyst has either got to pass the file to someone else, or change his tune. What do you think he's going to do?"

That's all hindsight. Stanley got caught up in the hysteria. Instead of asking for proof that Busang was real, not just a symptom of wishful thinking, an upbeat market and insatiable investment bankers, he dived in, feeling lucky to buy at $17 in the autumn of 1995. He borrowed heavily, more than $1 million, sinking it all into the stock. Loans were easy to come by; everyone figured it was only a matter of time before Bre-X partnered up with a major mining concern. The Street was hoping for a straight share swap. That way, Bre-X shareholders would end up holding stock in a major mining company, a big reward that precluded paying taxes in the short term, as no cash was expected to changed hands. "We were all looking for that tax-free roll-over," Stanley says. "Everyone wanted it. Walsh held it out for all his loyal shareholders. He'd talk about how Diamond Fields did it with Inco."

Walsh was turning into a cult figure on Bay Street, promising all his followers that they would achieve great fortune with him. Analysts quickly developed a loyalty towards the stubby, rumpled, chain-smoking promoter. "I think everyone, at least once, must have wondered if Walsh even had what it took to run a grocery store, let alone a mining company," says Stanley. "But it wasn't until much later that anyone checked his background. All you ever heard was that he'd been bankrupt. People were high on him because he controlled Bre-X. That was it. In person, he really didn't inspire much confidence."

Stanley began to have serious doubts about Walsh's credibility in early March 1996, following an annual mining conference in

Toronto. Bre-X was scheduled to hold its own annual meeting at the close of the event. It would be the company's first shareholder gathering ever held outside Calgary, primarily for the benefit of Bay Street. By this point, Bre-X stock was trading at more than $150, and Walsh, Felderhof and the rest of the Bre-X gang were being treated like celebrities. More than five hundred shareholders gathered in the ballroom of the Royal York Hotel, waiting for the meeting to start. When Walsh and Felderhof finally appeared, smiling and glad-handing like a pair of politicians, they were greeted with a standing ovation. Stanley stood at the back of the room and watched.

Felderhof spoke first. The impatient, craggy-faced geologist lectured the crowd about late dome volcanism, sulphide assemblages and multiple-stage hydrofracturing. He talked; the crowd squinted and tried to follow. Finally came the pay-off, something that everyone could understand: "All I can say is that we have thirty million ounces, plus, plus, plus." Mining Busang would be a breeze. All you had to do was scoop earth out of the ground and shake. Most of the gold would simply fall out, like pennies from heaven.

Then a stumble. Walsh took the podium and announced he was postponing the vote on a number of motions on the agenda, including a ten-for-one stock split and a shareholder rights plan. They would be considered back in Calgary, at a later date. Shareholders could use the extra time to think about the proposals, Walsh explained. Stanley thought this was unusual. So did a number of other shareholders. They examined the company's articles of incorporation to see if Bre-X had the right to postpone voting, and instead discovered that the charter prevented it from holding a shareholder meeting outside Alberta. Walsh didn't have the shareholders' interests in mind at all. He had screwed up and was afraid to admit it.

"Instead of coming clean, Walsh lied to us," Stanley says. "I just found that unnecessary and, you know, pretty strange." Stanley is loosening up. I get the feeling that this is a kind of therapy for him, getting things off his chest.

Troubling stories began to swirl around Bre-X, he says. People were whispering about its technical work, about its ownership of

Busang and about its enemies hovering in the wings. Stanley knew that games were being played in high places; there always are when a small company attracts the attention of the heavy hitters on Bay Street. Successful junior exploration outfits can be vulnerable prey for larger firms. Predators spread debilitating rumours, knocking down their target's market value before moving in for the takeover. It's ugly, and immoral, and it happens all the time.

In this particular case, Stanley wasn't sure what to think. He didn't know if there was any truth to the rumours that Bre-X was being targetted by bigger interests. There was. Information travels quickly in the mining community, even in the outback. Any geologist with a briefcase-sized satellite phone and an Internet connection can swap stories. And in 1996, the hottest topic on everybody's mind was Bre-X. James MacDonald, an exploration manager with Homestake Mining Company, was working high in the Andes when he heard the first rumblings of discontent via e-mail. A consultant friend of his in England labelled Busang as "most questionable." Others, based in Australia, "fingered the scam before it became obvious," MacDonald told me.

In Canada, a mining consultant named Dale Hendrick was also tossing dirt all over Bre-X. "He told me that Busang was a scam," says Stanley. "He said he had contacts in Australia who had drilled the site back in the eighties and hadn't found anything. He told me how unscrupulous mining companies sometimes drop bits of gold in their drill samples to make it look like they have found a rich deposit. He was basically saying that Bre-X had salted its drill core."

The speculation stemmed from the unorthodox manner in which Bre-X was handling its core. Usually, an exploration company gathers these cylindrical rock samples in two- or three-metre sections and then slices them down the middle. Half is crushed into a powder and assayed — or tested — for gold. The other half is put in safe storage and can be used to back up a company's initial assay results in the event that more evidence is required. But Bre-X wasn't splitting the core that it drilled from Busang. Instead, it was crushing almost all of it, save for a ten-centimetre-long "skeleton" sample.

This was a highly unusual practice for a small mining company still in the exploration phase. There was no way its staggering drill hole results could be double-checked by a potential partner. What's more, the company's assay lab in Indonesia was using a cyanide solution to chemically leach the gold from the crushed core. This also raised eyebrows, because companies generally use powerful blasts of heat to do the job in early stages of exploration. So-called fire assays are considered much more accurate and require smaller samples.

An article appeared in *The Financial Post*, drawing attention to Bre-X's use of cyanide. For some reason, the paper downplayed the crushed core angle, which was far more troubling. In any event, Bre-X took the matter seriously. John Felderhof lashed back in a press release, mocking those who dared question his decisions, telling them to "go back to school," that he did not "have the time to educate them."

Felderhof's arrogant response did nothing to soothe Stanley's concerns. He thought about unloading his shares, until he caught wind that Barrick Gold Corp., led by the irascible magnate Peter Munk, was preparing a hostile takeover of Bre-X. This made Stanley wonder if Dale Hendrick was a Barrick agent, hired to spread nasty stories about Bre-X in order to drive down its share price prior to a market takeout. Stanley knew enough about corporate warfare to consider it a possibility. He didn't know whether to trust Hendrick. "His opinion would change," Stanley says. "Sometimes he thought Bre-X was a scam, sometimes he thought it had gold. So I just let it ride."

In June 1996, new reports began to swirl on Bay Street. This time, word was that Bre-X's ownership of Busang was in dispute. Michael Fowler, a metals analyst with Levesque Beaubien Geoffrion Inc., noted in a memo that "this rumour relates to BXM [Bre-X's stock ticker symbol] either losing land title or some Indonesian partner gaining 25% of the property for a value much lower than the market." This time, the story didn't make its way into the daily press for another four months, long enough for an intricate back-room campaign to take root and grow into a conspiracy of major proportions.

Stanley was almost convinced there was something wrong; he just wasn't sure what it was. He was getting no help from Bre-X. The company's young investment relations team simply parroted the party line that nothing was happening, everything was in order, new drilling results would be coming out soon.

In Stanley's mind, David Walsh was delivering an abysmal performance as the primary custodian of a $6-billion investment. The least he could do was try to show some leadership, Stanley thought. Instead, he disappeared. On the rare occasions when he did field questions about Busang, he would reel off a glib statement, usually from a prepared text. Stanley's success as a special situations analyst depended on consistent, reliable information. Nothing was coming from Bre-X. He realized that he would have to sort the rumours out himself. He would start by checking out Busang.

But this presented another problem. Bre-X didn't want him there. The company never allowed outsiders to visit its operations in Indonesia, a fact that had vexed journalists and investors alike, although it hadn't dampened their enthusiasm for Bre-X. A little mystery often helps sell a story. Geologists from neighbouring sites were also prevented from setting foot on Busang, a stunning breach of mining etiquette. Site exchanges are common in the industry. Geologists treat their discoveries like children and love to show them off. As scientists, good geologists require proof of any alleged discovery. Swapping technical information regarding any geological achievement is almost seen as a duty and is certainly considered a pleasure. Bre-X's unfriendliness seemed strange, but it was shrugged off as one of Felderhof's idiosyncrasies.

On the other hand, Bre-X loved to show off Busang to friendly gold analysts. Stanley knew that the company was arranging a couple of major visits for North America's highest-profile analysts, part of an attempt to buff the company's image before negotiating a joint-venture deal. Since his specialty was outside of mining, Stanley wasn't considered for the tour. "I asked David Walsh's son Brett what my chances were of getting to Busang on one of those trips. He said, 'Zilch. We only want gold analysts on the property.'"

Stanley decided to make the eighteen-hour flight to Jakarta anyway and try to meet with some of the almost three dozen other Canadian exploration companies that had set up operations in Kalimantan. Indonesian proximity plays were almost as hot as Bre-X. In late 1995 and early 1996, obscure penny-stock outfits were scooping up tens of millions of dollars through private placements and initial offerings, merely by staking mineral claims close to Busang. Stanley had been investing in a handful of them, betting that they would rise with Bre-X.

Lacking any expertise in geology, Stanley was focused on political risk. He'd been to Indonesia before and knew the way in which President Suharto ran roughshod over everything. Security of tenure and legal title, distinctly Western concepts, mean little in autocratic Indonesia, where the success of any venture depends more on political connections than business acumen. Suharto and his close band of cronies hand favours to faithful acolytes like little candies and snatch them away the instant anyone displays any independence. His minister of Mines and Energy, Ida Bagus Sudjana, was known for putting the big squeeze on foreign companies, demanding equity stakes in major Indonesian projects from players such as Denver-based Newmont Gold Corp., one of the world's largest gold producers. Had Bre-X and the other Canadian exploration companies developed good relations with the appropriate people? Were their properties safe from predators? These questions hadn't even been raised in Canada, an astonishing fact, given that the combined market capitalization of Canadian exploration companies in Indonesia exceeded $8 billion. That was a lot of money hanging on the mood of one foreign dictator.

"No one had bothered to check on the Indonesian end," says Stanley, shaking his head and taking another pull on his beer. Prior to his trip, he had called on a couple of Bay Street analysts to get their views on Indonesia. No one could tell him anything. He dropped by the offices of a junior mining company with some freshly acquired properties in Kalimantan. "The president didn't have a clue about Indonesia," he recalls. "I asked her where her property was. She wasn't exactly sure, except that it was near Busang. She was totally ignorant about Indonesian politics and business. I don't think she even knew who Suharto was." The market didn't care. Earlier in the year, the company had raised $10 million in a private placement.

In early July, Stanley landed in Jakarta. He hit the ground running, visiting three of the Canadian companies in which he'd invested, establishing an information pipeline that later came in very handy. He also met a number of political analysts. One of them laughed when Stanley asked if Bre-X's ownership of Busang was secure. "Why should it be?" he snorted. "This is Indonesia. Suharto giveth, and he taketh away." That's all it took for Stanley to start worrying that Bre-X might be vulnerable to some kind of covert power play. He had no idea there was already one in motion.

More determined than ever to get to Busang, Stanley invited Bre-X's commercial manager to lunch and got him to check if there was any more space on a chopper carrying a group of analysts to the site. He got lucky. George Milton, a stockbroker with RBC Dominion Securities Inc. and an early Bre-X proponent, had been forced to cancel his trip to Indonesia. There was one extra spot on the helicopter, and Stanley could have it if he could get to Kalimantan on time. "I could hardly sleep the night before I left," he says. "I was getting a chance to see the most sought-after gold property in the world."

On July 14, he grabbed the four-hour flight to Balikpapan, an oil and gas centre and the largest city on Borneo's southeast coast. He checked in at the Dusit Inn, a luxurious business hotel, the best in Kalimantan, surrounded by lush, tropical gardens, and a favourite with the Bre-X management. Soon he was shaking hands with seven gold analysts who had been formally invited to inspect Busang. The line-up included a pair of Wall Street heavyweights, David Neuhaus, from J.P. Morgan & Co. Inc., and Daniel McConvey, a former Barrick Gold employee representing Lehman Brothers, Inc. Stanley also noticed a small, bespectacled woman named Vivian Danielson, a writer with *The Northern Miner*. Danielson had been one of the first journalists to give credence to Busang, and her writing had brought Bre-X to the attention of some of the world's biggest mining companies. For that, Walsh was extremely grateful. Rewarding Danielson with a site visit was his way of tipping his hat and keeping their happy relationship intact. Six months later, *The Northern Miner* distinguished Felderhof and Walsh as its "Mining Men of the Year," for identifying "one of the most important gold discoveries of the century."

Stanley realized he was with an influential bunch of people. If there was any problem with the Busang deposit, this gang would spot it. He assumed they were as inquisitive as he was. He was wrong.

The first night in Balikpapan, everyone gathered at the bar in the Dusit Inn. Felderhof was there, and de Guzman. There was plenty of drinking. Eventually, some of the analysts went off to inspect the local nightclubs — booze cans, basically, where lonely expats are expected to belch karaoke and grab at the flirty bar girls. Stanley stayed behind at the hotel and ended up playing a game of pool with John Irvin, manager of Indo Assay Laboratories, the man in charge of testing Bre-X's drill core. "We were sort of joking around," Stanley says. "I asked him if there was really any gold up at Busang. He wouldn't look at me. He mutters that there might be some problems when other people got around to testing the core. He said they might have trouble reproducing his results. When I started to ask Irvin what he meant, he made up some excuse and left."

Stanley let it go and went to bed. Early next morning, he joined the rest of his party at the Balikpapan airport. Some of the analysts looked a little green as they boarded one of two assigned helicopters waiting on the tarmac. De Guzman, Felderhof and Felderhof's nine-year-old son, John Junior, climbed aboard the smaller chopper, a French-built Alouette III. Stanley glanced at the Alouette's pilot, a diminutive, middle-aged Indonesian named Edy Tursono. He noticed Tursono was wearing a large ring on his left hand, featuring a skull and crossbones.

Busang is about 250 kilometres inland from Balikpapan, a distance covered in a little over thirty minutes via helicopter. As was the usual practice, the Bre-X aircraft touched down briefly in Samarinda, another major centre north of Balikpapan, where Bre-X stored most of its core before sending it to the lab. Back in the air, Stanley gazed down at the wide swathes of second- and third-growth forest carpeting the ground below. It was mid-morning by the time the two helicopters landed at the main camp. The analysts were assigned beds in a small, tidy sleep house and issued bath kits. They were then driven around the property in a fleet of jeeps, stopping frequently to scramble down a river bank to look at outcroppings of rocks. It was hot, humid and, in Stanley's view, exasperating.

"Felderhof would explain the significance of this rock, that pattern, while de Guzman stood beside him wearing his ball cap and sunglasses, grinning and nodding his head, and then we'd be off to the next spot," Stanley recalls. "Everyone was very anxious to see the geology and take notes, but none of the guys were asking any tough questions about the drilling and assaying. Maybe they were afraid they'd get an earful."

He decided to toss out an obvious one. If he was not going to split its core, Stanley asked Felderhof, why didn't he at least drill twin holes, side by side, and save one set of core as back-up? Surely this would ease any worries back in Canada. Felderhof glared at Stanley. A few moments of uneasy silence passed. "I thought he might tell me to get the hell off his property," laughs Stanley. "Then he said that twin holes would be too expensive, and that all the necessary precautions had been taken. He said they might twin in the fall. End of story."

Again, none of the gold analysts had anything to add. They weren't interested in assessing Bre-X's technical methodology. It appeared to Stanley that they were more intent on covering Egizio Bianchini's tracks. The senior gold analyst from Nesbitt Burns had already been to Busang twice. A few weeks earlier, in fact, he'd taken a private tour with a major client, American mutual fund giant Fidelity Investments. Fidelity was so impressed with Busang it snapped up fifty million Bre-X shares, more than seven per cent of the company's total float. It was easily the largest single retail position in Bre-X, and it won Nesbitt Burns a big-league commission. That was the action the other brokerages were clamouring for. And all the analysts knew it.

When they arrived at the so-called discovery outcrop, de Guzman grabbed centre stage, installing himself in front of a patch of loose, weathered rock. Chubby, self-absorbed, yet totally deferential to Felderhof, de Guzman, forty, didn't cut a very authoritarian figure, but it hardly mattered. His audience was captivated. This was what the analysts had come for. What they were presented with looked rather ordinary. De Guzman ordered them to look closely. The reddish-brown sedimentary rock, sitting on top of the black, carbonaceous rock, that's a clue, he said. There's gold in there, but it's about ten metres down. It's all leached out at the surface. That's why the dozen or so other mining companies that walked through and took samples never found anything, de Guzman told the analysts. Bruno Kaiser, metals analyst from CIBC Wood Gundy, collected some small rocks off the surface and shoved them in his pocket.

They drove past some drilling rigs. Stanley noted that there were about half a dozen Filipino geologists, all of whom answered to de Guzman. They were firmly in control of their six drill crews, made up mostly of Australians, distinguishable by their distinctive outback hats. "The Filipinos were working the drill crews like dogs," Stanley says, adding that he felt reassured by this. "To me, it meant that the intensity level was high, that this was no laid-back summer camp."

The group arrived back at the main site, tired and hungry. They were revived with barbecued steaks and lots of Bintang, a ubiquitous Indonesian lager. They'd put in a full day, but most of the analysts stayed up late, drinking with Felderhof and de Guzman and listening to them bitch about life in the jungle.

"Felderhof was lamenting that he didn't get the recognition he deserved," Stanley recalls. "He thought that Walsh was getting most of the credit, while he was doing all the work in Indonesia.

De Guzman was doing his nodding and grinning routine again." De Guzman made himself out to be an intrepid adventurer. Some of his stories were hard to believe. "He told us that, years earlier, he'd been doing some really rough exploration work for a mining company in the north, near Sarawak. It was clandestine stuff, and he had to be dropped off in the jungle by military helicopter. He had to find his own way out and ran into a tribe of headhunters, who demanded that he give them his gold watch. He was proud to tell us that he made it out of there in one piece, wearing his watch."

De Guzman also described his fear for his family back in the Philippines. People with money had to be wary of kidnappers, he said. Someone, maybe a rival mining company, had hired private investigators to check out his background. They were hounding his relatives. De Guzman said he and his close Filipino colleague, Cesar Puspos, were thinking of moving their families down to Perth, in Western Australia, where they could live in peace.

Stanley fell into his bunk, exhausted. The next morning was spent flying over the Busang property. Back on the ground, a couple of analysts began pestering Felderhof for some core samples to take home, souvenirs. Felderhof grimaced and then took de Guzman aside. The two men talked briefly, and de Guzman disappeared. He came back with some pieces of cylindrical rock, dark brown, about six centimetres thick, skeleton core from one of the holes. This was broken up and handed out.

Finally, the analysts were assembled in a small map room. Felderhof reviewed the spectacular drilling results from the south-east zone. He pointed to a map, crossed with lines. Each line was numbered and represented a distance of about one kilometre. There were anywhere from two to twenty black dots on each line. These were the drill holes, he explained, some of them five hundred feet deep. The latest holes were turning up incredible grades of gold, Felderhof announced. He led them through the math: Take the average grade per hole, measured by the assay lab in Balikpapan, multiply by the area drilled, and ka-*ching*, there was your estimate, endorsed by PT Kilborn Pakar Rekayasa, a division of the Montreal-based engineering giant SNC-Lavalin Inc. It was beautiful arithmetic, more meaningful to that group of analysts than any mundane piece of rock from the jungle.

For the finale, Felderhof pointed to a distant drill hole, sitting way out there, all alone. It was labelled LBH1. This was a very special hole, Felderhof crooned. This hole would blow everyone away. It was showing three grams of gold per tonne of rock, a decent grade, to be sure. But the real thrill was that it sat one and a half kilometres to the northeast of the closest drill line, number 69. Felderhof was suggesting that the grade ran consistent all the way through the gap. The most recent published estimates for the south-east zone identified thirty-four million ounces, but he was confiding that there was much more. The gold deposit went on, and on, and on. It was bigger than even he could imagine. De Guzman nodded and smiled.

The analysts started scratching away on their notepads, factoring in the additional land mass between the last fence line and LBH1. Their calculations varied between 200 million and 400 million ounces. "I can't write that," Stanley heard one of them gasp. Who would believe it? The world's largest known gold deposit, discovered by Freeport-McMoRan Copper & Gold Inc. eight years earlier in the remote Indonesian province of Irian Jaya, contained fifty-two million ounces. Other major mining companies such as Barrick had total proven reserves of fifty million ounces or less, from myriad properties scattered around the planet. If Felderhof's supposition was correct, Busang might contain more gold than all the known deposits in the world combined.

The satellite link back to North America was busy all night as the analysts took turns faxing in their early reports. Within days, they were being devoured by investors from coast to coast. Every report recommended Bre-X as a strong buy. Most of the analysts on the trip couldn't resist spreading word that Kilborn's new forty-seven-million-ounce estimate was just the tip of the iceberg. The suggestion that Bre-X might be hyping its estimates was not even entertained. This baby was real.

- "Arriving at a final gold resource number for Busang is difficult since the deposit is beyond the scope of anything seen to
date. . . . Geological evidence is very strong that they will meet
or exceed [fifty million ounces by the end of 1996]. . . . A visit
to the property is necessary to grasp the magnitude of what is
going on and to justify throwing out the rule book."
  – Bob Sibthorpe, Yorkton Securities Inc.

- "The ease with which the company is adding to resources only
adds to the cache of this amazing deposit. . . . We expect that
exploration at Busang will continue to add significantly to the
level of resources as the deposit remains open in most directions."
  – Jim Taylor, Yorkton Securities Inc.

- "In our opinion, the company is conducting a very professional
exploration campaign. . . . Activities are systematically executed
in a way that can be easily verified."
  – Chad Williams, P.Eng., Research Capital Corporation.

Egizio Bianchini at Nesbitt Burns weighed in from his privileged
vantage point, interpreting the new Kilborn results as "further confirmation that the Busang deposit is among the two or three largest
gold deposits in the world. . . . Much more mineralization waits to be
discovered." The common consensus had zoomed to around seventy
million ounces. Back in his office in Manhattan, J.P. Morgan's David
Neuhaus made the boldest prediction. "I'd say 150 million ounces is
a conservative guess as to what Bre-X will ultimately come up with"
at Busang, Neuhaus told *The Financial Post*. Two months later, J.P.
Morgan was named as one of Bre-X's new "financial advisors."

*Further Confirmation. Geological evidence is very strong. Easily
verified. Conservative.* The analysts were promising Bre-X had gold.
These were extraordinary comments. None of the analysts had seen
a grain of gold from Busang. Nothing had been verified. As Bob
Sibthorpe had suggested, they were *throwing out the rule book*.

Bill Stanley read the reports when he got back to Toronto. What he read didn't surprise him; he'd anticipated that his colleagues would move their estimates up after learning about LBH1. But Stanley remembered his encounter with John Irvin, the nervous assayist. He recalled how John Felderhof had ducked his question about twin holing. And then he remembered his core sample.

He called around to see if the other analysts had tested their portions for gold. No one had. Stanley had his assayed towards the end of August. "I shared the results with a number of the guys," he says. "They told me it meant nothing. They said the sample was too small to be relevant." Still, he would have been relieved had the test turned out positive. "There was nothing there," he says. "Not a thing."

3

GOLD FEVER

Next to horse-dealing, I suppose that the purchase and sale of gold mines has been associated with more fraud than any other commercial operation.
– Charles Dobson, in *The Cosmopolitan*, April 1898

Regretfully I must confess to a feeling of cynicism and despair for the human race as a whole.
– Canadian geologist Franc Joubin, 1986, after five decades in the mining industry

"**ALL THAT GLITTERS IS NOT GOLD,**" wrote William Shakespeare, but humans have never heeded the bard's sage warning. Gold's seductive gleam has been praised and lamented for six millennia, by Pharaohs, philosophers, poets. It has been worshipped and mythologized, hammered into idols and shrines and beamed back to the heavens in hopes of attracting God's favour. Gold is useful, eternal, impervious to the elements, and beautiful, an earthly embodiment of truth and power. No wonder that it has always been hoarded by governments. No wonder, then, that it's the favourite commodity of charlatans, gangsters and crooks.

As much as man has rejoiced over the discovery of gold, he's also regretted it. "I hate gold," sneered the ancient playwright, Plautus. "It has often impelled many people to many wrong acts." Another Greek wrote that "gold and silver are injurious to mortals. Gold is the source of crime, the plague of life, and the ruin of all things. Would that thou were not such an attentive scourge! Because of thee arise robberies, homicides, warfare, brothers are maddened against brothers, and children against parents."

The search for gold has always led to jealousy and greed and other profanities. The first gold mines were tunnelled by slaves. Three thousand years ago, Egyptians bound their captives and threw them into mine shafts, where they spent their lives digging. Diodorus, a Greek historian, wrote that "living in darkness, they contort their bodies this way and that to match the behaviour of the rock. . . . There is absolutely no consideration nor relaxation for the sick or maimed, for aged men or weak woman; all are forced to labour at their tasks until they die, worn out by misery, amid their toil."

By the Renaissance, it was noted that mining had led to environmental destruction. In 1555, the German scholar Georgius Agricola advised that "there is need of an endless amount of wood for timbers, machines, and the smelting of metals. And when the woods and groves are felled, then are exterminated the beasts and birds, very many of which furnish a pleasant and agreeable food for man. Further, when the ores are washed, the water which has been used poisons the brooks and streams." But Agricola, a humanist and scientist, argued that "those who speak ill of metals and refuse to make use of them do not see that they accuse and condemn as wicked the Creator Himself." His comments could easily have been written today.

Mystics have assigned gold magical qualities; others have prescribed it as a treatment for ringworm, ulcers and piles. Alchemists concocted recipes to turn base metals into gold and gave us brass instead. A British friar and empirical scholar, Roger Bacon, defended his attempts to meddle with nature, writing that "the search and endeavours to make gold have brought many useful inventions and instructive experiments to light." But gold itself is a native element; it can never be conjured, only found or stolen.

Spanish conquistadors set off to the New World and seized tons of gold belonging to the Incas and Aztecs. They also built mines with slave labour. Offended by the conquistadors' profane greed, native warriors poured molten gold down the throats of captured soldiers. But the thirst was unquenched; Europe screamed for gold leaf, gold jewellery, gold threads for their finery. Convoys, organized and financed by the Spanish Crown, began sailing twice a year from Seville to the Americas, returning with 750,000 pounds by the end of the sixteenth century. Spain dominated the industry, from production, to transport, to trade, and her regal convoys attracted all kinds of poachers and pirates. Sir Francis Drake became a legend after his rampage off South America's Pacific coast, where he looted dozens of Spanish ships and brought home at least twenty tons of gold.

There were other opportunists. As the traffic in gold became bigger and farther reaching, so did the investment frauds. In his famous mining text *De Re Metallica*, Agricola described how promoters "advertise the veins with false and imaginary praises, so that they can sell the shares in the mines at one-half more than they are worth."

It is fitting that one of the earliest gold swindles was born in a desolate, windswept territory that would later be known as Canada. In 1577, Martin Frobisher, a British naval explorer, returned to England after one of many failed attempts to locate a northwest passage to China. Frobisher's journey had stalled in the Arctic Ocean, off the coast of Baffin Island; freezing temperatures and dwindling provisions forced the commander and his fleet of three ships to sail home. Before turning around, however, Frobisher grabbed a number of souvenirs, including a local human specimen and a large, black stone, snatched from the shore.

Once back in England, Frobisher's stone fell into the hands of an Italian alchemist named Giovanni Battista Agnello. This was no worthless rock, Agnello proclaimed. This was a gold-bearing mineral complex that, he boldly predicted, was capable of producing twenty-five ounces of gold per ton. Frobisher wasted no time putting together another expedition, joining forces with a local merchant named Michael Lok, who served as his promoter. The plan was to return to Baffin Island and load up with as much of the fabled rock as possible. Lok had little trouble raising money through a public offering. The scheme was backed by Queen Elizabeth I, who dropped £1,000 into the pot. Frobisher returned with 158 tons of black stone, which, it was advertised, was even richer in gold than the first small sample. Investors were told to remain patient while assayers figured out how to remove all that gold from the rock. A third expedition was launched and this time, Frobisher controlled four hundred men and a fleet of fifteen ships. He sailed home with more than fifteen hundred tons of stone. Once again, a team of assayers set about to divine the precious metal.

They found nothing. The rocks were worthless. Lok stalled for more time, protesting that the assayer's methods were faulty. More tests concluded that there was more gold in common sea water than in Frobisher's stones. Lok declared bankruptcy and was sent to debtors' prison. Frobisher avoided prosecution by claiming to have been an unwitting victim. It worked; he became a vice-admiral under Drake and fought the Spanish Armada. It's believed that his "gold" was in fact amphibolite, a dark rock with specks of shiny mica. A few trenches left over from his expeditions were discovered on one isolated island in the bay that now honours his name.

Frobisher's sorry experience was largely forgotten, lost in the archives of time. Real gold discoveries in Africa and Australia grabbed everyone's attention. Later explorers to Canada, including Samuel de Champlain, received gifts of copper from local natives and heard stories of large mineral deposits, but most were too busy charting territory, making political pacts and establishing a lucrative fur trade to bother with the territory's mineral potential. A hundred years would pass before the first ore deposits were mined in Quebec and converted to iron, and another century before coal and copper would come into production. After the formation of the Geological Survey of Canada in 1842, explorers finally had the benefit of comprehensive maps and geological reports. Engineering and mining schools were established, and a new professional industry was born. Still, there has always been a place for rank amateurs.

Whispers of gold first hit Canada's rugged west coast in 1857, just after the California gold rush. Prospectors got word that the Fraser River in British Columbia was panning well; thousands stampeded north to check out the scene and indeed, there was gold, but not enough to sustain all of them. They pressed farther north, to the mountainous Cariboo region, and found some more. A handful of mines were opened in the British Columbia interior, in places such as Nelson, and Trail, and the Kootenays. Then came the Klondike.

Unlike earlier bonanzas, which were managed largely by royal or corporate concerns, the Yukon gold rush was a free-for-all, with few regulations and restrictions. As Douglas Fetherling notes in his book *The Gold Crusades — A Social History of Gold Rushes,* most of the thirty-five thousand men and women who made the epic journey up the British Columbia coast, over the dreaded Chilcoot Pass and into the Yukon frontier were ordinary, restless people, seeking some kind of fulfilment and connected by a "rootlessness born of optimism, or else the disillusionment that they perhaps blamed on conditions at home or imperfections of the age."

About $100 million worth of gold was picked from the Klondike, billions of dollars by today's measure. It was mined mostly by journeymen who dredged tiny flakes from the gravel and sand beds around Dawson City. Their haul was a pittance compared to the $2.5 billion now mined in Canada annually, but, as Fetherling points out, it was "the journey and the process, not the destination or the fact," that seeped into a new nation's psyche. Mining introduced a thrilling entrepreneurial spirit to Canada, one that spread from the traditional centres of power such as Montreal and Toronto, where exploration capital was raised, and into the hinterland, where the money was spent.

Huge mineral discoveries along the Canadian Shield, a vast belt of Precambrian rock stretching from Labrador to the Northwest Territories, became almost routine in the early twentieth century, creating thousands of jobs and making millionaires out of ordinary men. Without Canada's immense mineral production — including an annual allotment of two billion pounds of nonferrous metals — to back the Allied effort, the Second World War would have dragged on far longer than it did. The two atomic bombs that fell over Japan, after all, contained uranium mined from the Canadian north. By 1945, Canada was the world's leading producer of nickel, platinum, asbestos, iridium and paladium; second in zinc and cadmium; and third in gold, copper, magnesium and cobalt.

New exploration techniques, including airborne reconnaissance and magnetic surveying, helped further the mining industry's ascendancy in the post-war years. With every innovation came more discoveries, more geologists and more investment. By the 1990s, Canada was experiencing a full-fledged mining boom. Thanks to new open-border policies in former communist countries and isolationist states, Canada's mining frontier spread to every corner of the earth, to countries such as Albania, Kazakhstan, Mongolia, Burkina Faso and Chile. There was no shortage of money flowing in to fuel the expansion. In 1996, Canadian miners raised US$6.5 billion on the stock market, more than all the mining companies in Australia, the United States and South Africa combined. Why? Their stock was cheap, their companies small and portable, and they didn't carry much debt. When they did hit, the pay-off was spectacular.

When the price of gold drops — as it did in late 1997, to its lowest level in decades — new mining ventures seem about as attractive as a Saskatchewan time-share. Even established producers such as Barrick Gold and Placer Dome Inc. become vulnerable. Both companies were forced to close mines and eliminate jobs when gold dropped below US$300 an ounce, the benchmark price for profitability.

But the industry never fades away, it just rests a while, waiting for the inevitable rebound. Wars, drought, recession, anything that can cause economic uncertainty help move gold prices higher. There are over two thousand publicly traded mining companies in Canada. Only 150 or so actually own producing mines inside the country. Of the rest, most are two-bit exploration outfits with no prospects, only smooth-talking promoters and optimistic geologists. They stand a one-in-a-hundred chance of finding something, but persuasive shills can always attract money. Gold's lure is etched in the national psyche. It never fades.

Desperate promoters, lying geologists, phoney assays, mysterious deaths: the business is full of war stories. In 1934, the magazine *Canadian Forum* threw cold water on the modern art of stock promotion, calling the "current deluge of gold mining issues" a "racket" that was "by nature dishonest, in practice outrageously extravagant, and in result disastrous not only to the investing public but to the mining industry itself."

Entire books have been written describing how investors are gleefully bilked out of their savings by winking promoters. The industry loves to recount how mischievous Murray diddled them on Bay Street and screwed them on Vancouver's Howe Street. In 1967, a bespectacled public relations manager and former mining shill named Ivan Shaffer wrote an explicit and funny account of how unscrupulous promoters do their business. The mining business is "larcenous," he concluded, in *The Stock Promotion Business: The Inside Story of Canadian Mining Deals and the People Behind Them.* "I have never met an honest successful mining promoter.... A promoter's profit comes from the sale of stock. Promoters make money. They do not make mines except by accident."

Bogus mining stunts are as Canadian as hockey sticks. Investors still haven't learned to spot the difference between an honest promotion and a scam. And no one has wanted to teach them. That would ruin everything. The best deceptions are ridiculously simple and rubbery enough so that blame will bounce. They involve just a handful of players. Sometimes only one is in on the scam. The longer the deception continues, the more people join in, becoming actively involved or turning a blind eye. Without others, the scam can't roll through the whole process.

At the top of the pyramid are the fast-dialling promoter and the freelance geologist. Brawn and brains, and mutual dependency. The promoter, battle-scarred but potentially profitable, has the confidence of the regulators and the junior stock exchange, which is always looking for new listings to add to the stable. All it takes to obtain a listing on the Alberta Stock Exchange, a popular promotional venue where Bre-X was born, is a loose prospectus and some cash. No detailed business plan is required, just a promise to do something.

The geologist, a hard-bitten, opinionated figure, has one or two moldy university degrees. Any youthful fantasies of finding eldorado have been superseded by a basic instinct to survive. He needs work. He has no money, and no luck, but he does have a line about a nice little property out in the bush. It's well-removed, hard to get to, with plenty of mineral potential. The geology is anomalous, of course, not conventional. That's why it has been overlooked. The best stories are always a little outside of convention. A dab of mystery is key to the show.

So the promoter starts extolling the hidden charms of his new property. He doesn't have any drilling data yet, but his "independent consultant," the geologist, has inspected the site and is convinced there's gold there. Why not take a punt, the promoter coos to friends, neighbours and associates. Get in now, when the stock is cheap, because this baby is going up. Important issues such as ownership, recovery rates and construction costs are either ignored or put on the shelf for later discussion. What matters most is the immediate pay-off.

The promoter hires people to help him sell the message. He sets up a boiler room, filling it with young men and women fresh out of school, armed with telephones and a list of phone numbers. They may even cold call, trolling for mom-and-pop investors. The promoter concentrates on the analysts and investment "professionals" who work for the banks and the brokerages.

He calls and he calls, and if he's got a good pitch, a good story, the pros have a look. They jump in knowing that, if nothing else, they can at least move the stock. If it's a very good story, with some hot exploration "results," the pros might attract the big buyers, the institutional investors and mutual fund managers with real money at their disposal. At the bottom of the pyramid are the retail investors, tormented figures who, like the other actors, are motivated by profit. But information doesn't always trickle down to them. They don't know much, but they expect their money to grow. They've seen the ads on television. They're in good hands.

There is always some scepticism at some early point in the promotion, but it's minimized with an endorsement from an authoritative character, a former Bay Street executive, now retired, brought on board with cheap shares and options. Drilling commences. Samples are assayed and found to contain encouraging quantities of gold (or silver, or nickel, etc.). The promoter builds a wall around his deposit. No one is allowed to cross onto the site. The stock moves up and down, fuelled by gossip leaking from company headquarters. Money changes hands among the middlemen. The Pros file reports.

Rumours that the mineralization may have been overstated are expressly denied by the people in charge. The "confirmed" size of the deposit is moved up. The market goes nuts, and the promoter and his friends make a pile of money exercising options, buying and selling shares on the open market or in secret. They may also make money selling the stock short, anticipating a collapse.

Ultimately, the fraud is exposed, usually when a larger, predatory company pushes its way onto the scene and gets burned. It comes to light that samples were tampered with, assay results fudged, information destroyed. This is always followed by denial. The promoter claims he is innocent. The geologist cannot be found. Sometimes, people die.

Investors, meanwhile, vow to never make the same mistake again. The small buyers demand justice, public floggings, a crackdown in the mining industry. They are ignored. Institutional buyers, the brokerages and mutual fund companies, are silent. Their analysts,

the "geological experts" who "dig deeper" and are supposed to analyse every facet of the promotion before recommending the stock,
shrug their shoulders and move on. Not our fault, they invariably
say. Memories fade. The cycle repeats itself with new players, using
the same methods, obtaining the same results. Gullibility and greed
cannot be destroyed. This is not cynicism or post–Bre-X wisdom.
This is history.

Most mining swindles are merely variations on the Frobisher
theme, based on nothing more than a single, delicious lie. In 1964,
George and Viola MacMillan, a well-known husband-and-wife team
of Toronto mining promoters, bought a group of languishing claims
next to a massive copper discovery in northern Ontario. Then they
sat back and watched as their ordinary little proximity play quickly
turned into one of the most notorious — and effortless — mining
scams of the century.

Barely one day into their drilling program, rumours spread that
the MacMillans' company, Windfall Oil and Mines Ltd., had found
copper. The story presumably sprang from a few witless drillers who
had watched George furtively haul drill core away in his station
wagon. In fact, the core had been tested and was found to contain
no copper at all. But the MacMillans did nothing to stop the false
information from circulating; in fact, they sat on their assay results,
denying they had done any testing on the drill core. Meanwhile,
they were snapping up Windfall stock and selling as the market
responded in kind. The traders on Bay Street heard the rumours,
saw the activity and jumped in. The media followed along behind,
buzzing about the big copper find. Windfall shot from 56 cents to
almost $6 in little more than a week. Based on nothing.

Most investors don't know anything about mining. But the Mac-
Millans were well regarded inside mining circles. Viola, the "queen
bee of Canadian mining," was the president of the Prospectors and
Developers Association of Canada, a position she'd held for two
decades. This minor biographical note didn't mean the MacMillans
were trustworthy. But Windfall stock was hot. People didn't want to
miss out.

Windfall was an information black hole, sucking money in, letting nothing escape. The MacMillans played the game close to perfection, saying little, dodging requests for more news about the drill core. The couple netted $2 million before finally being forced to divulge their pathetic assay results. Windfall stock was crushed during trading the following day. The MacMillans were charged with manipulating the price of Windfall's stock, but they were never convicted. They rehabilitated their tarnished reputation by donating generously to various public causes and then got back to work, making deals. The affair's only lasting impact was an exodus of mining promoters, out of Toronto and into virgin territory on Canada's balmy west coast.

People searching for a closer antecedent to the Bre-X scandal need look no farther than Vancouver, the city that *Forbes* magazine famously dubbed the "scam capital of the world." Poison-penned journalist Joe Queenan, who wrote a rollicking account of the Vancouver Stock Exchange for *Forbes* in 1989, referred to a few of its most "ludicrous," "brain dead," "deliriously idiotic" mining promotions. Somehow, he overlooked the most sensational effort, a scandal that involved doctored drill samples and an unsolved murder. So did a lot of Bre-X investors.

New Cinch Uranium Ltd. was the brainchild of a down-and-out Vancouver mining promoter named Albert Applegath, and backed by Vance White, a Toronto financier. Late in 1979, the company announced it had stumbled on a gold and silver deposit in the New Mexico desert, a few miles from the Rio Grande. Several months later, New Cinch started to roll out some terrific results, from core assays conducted by an independent lab in El Paso, Texas. But a reporter from *The Northern Miner* who ventured down to see the New Cinch operation was puzzled to find no activity around the drill site, despite promises that rig workers were going full out. He did, however, encounter "three tough-looking hombres" guarding the core shack and brandishing "an assortment of sidearms and rifles."

New Cinch avoided scepticism by not reporting assay results that pointed to "negligible" amounts of gold and silver at its property.

Incredibly, upon review, the VSE allowed New Cinch to publish only the good numbers. The second, lower results were locked away in a confidential file, away from public scrutiny. The positive news kept coming. By November 1980, New Cinch was estimating that its "discovery" contained two million ounces of easily recovered gold. The company's stock soared from $1.21 to a high of $29.50.

Naturally, New Cinch attracted suitors. Willroy Mines Ltd., a Toronto-based mining company, decided to mount a takeover. "We bought stock based on the information New Cinch had published," recalls Ian Hamilton, a former director and legal counsel for Willroy, which later became Lac Minerals Ltd., which in turn was bought by Barrick Gold. "Our first mistake was not figuring out where the shares were coming from, who was selling them. The second mistake was that we continued to buy stock after they refused to let us on their property."

Willroy spent $26 million on New Cinch stock before it finally got on site and did its own drilling. The company shipped some mineral samples to assay labs in Vancouver and Toronto, and waited for the results. The assayers found nothing. "Naturally, we were wondering what the hell was going on," says Hamilton. "We didn't get any satisfactory answers. Applegath just gave us bafflegab. He kept saying something about the 'vulgarity of assaying.' That's when we knew we were definitely in trouble."

As a major holder of New Cinch stock, with new information about the company's property, Willroy had a duty to share its information with the public. As a precaution, more tests were conducted, with the same dismal result. There was only one logical explanation: New Cinch had faked its early assays. The stock cratered; Willroy sold out at 50 cents a share.

Salting drill core is the crudest, most contemptible form of mining fraud. No one involved in the practice would ever admit to it, of course, so it came as no surprise when Vance White came out swinging on behalf of New Cinch. His obstinate defence was conducted in a manner that Bre-X investors would easily recognize a decade and a half later. "There can be no doubt that gold and silver occur on the property," White insisted. It would have been "extraordinarily difficult" to salt its core over a period of two years, he added, and with such remarkably consistent results, without someone finding out. Preposterous. Impossible. Forget it.

A third round of independent testing destroyed any hope that the property contained gold. New Cinch was forced to admit that its drill core must have been tampered with. But White refused to accept any blame. He was, after all, a money man, not a technician. White pointed the finger at the company he hired to conduct the assays. New Cinch, he thundered, would "institute the necessary proceedings forthwith and pursue its remedies against those responsible."

Willroy figured it knew who was responsible, launching a $21-million lawsuit against White and New Cinch. The company was on its own, however; officials at the Vancouver Stock Exchange expressed the appropriate level of horror and went back to sleep. The Royal Canadian Mounted Police confirmed reports it was investigating possible criminal activity, but that was as much as they would ever comment on the affair.

Willroy hired its own private investigator to check into the scandal. The trail led him to Hong Kong, where, Hamilton says, the investigator was murdered. Local police treated it as a simple "misfortune" and closed the file. On 14 November, 1982, a former employee of the assay company New Cinch used was gunned down in his El Paso apartment. Three months before he was shot in the head, Michael Opp had told his mother he was afraid for his life, thanks to information he had relating to the bogus New Cinch assays.

Both deaths remain unsolved. The scam's perpetrators were never caught. The only ones who paid any penalty in the New Cinch affair were investors. Hamilton, who now runs his own junior mining company, says that the senior executives behind New Cinch "were, at the very least, guilty of willful blindness. It was incredible. They had a legal responsibility to ensure their information was the best. They hid behind these ridiculous excuses, and some people accepted that. It still happens. It just goes on and on.

And no one tries to solve the problem." The crooks, Hamilton says, are still out there.

New Cinch was peanuts compared to Bre-X. Canadians are now responsible for hatching the biggest scam in mining history, an artful, multibillion dollar fraud that endured for more than three years, involving some of the country's most powerful business interests. The deception has had a devastating impact on thousands of lives, in Canada and around the world. And yet there may be no recourse, no justice, no relief. In that sense, Bre-X was no different than most mining scams. Investors were lulled into making the same fatal mistake. They gave the wrong people the benefit of the doubt. Which leads us to David Walsh.

4

THE
PROMOTER

*When the palate revolts against the insipidness of rice
boiled with no other ingredients, we dream of fat, salt,
and spices.*
— Hindu expression

I've been up and down a couple of times.
— David Walsh

DAVID WALSH has seen plenty of trouble before.
Back when he was a nobody, he always managed
to keep it quiet. Now he can't turn around with-
out someone pointing an accusing finger at him. There are enough
lawsuits piled up against him to keep a truckload of lawyers busy for
years. People have threatened to blow up his office. His son was
charged with assault after allegedly attacking a television camera-
man. His wife sits alone in their garden estate in the Bahamas, trying
to recover from a nervous breakdown. Walsh says he feels foolish,
and angry. He insists he's an innocent victim of fraud. Others aren't
so sure.

Walsh is no neophyte in the mining industry. He is a classic penny-stock promoter. He's a gambler and a money hustler, with a long history of bad deals and busted relationships. He is corpulent and sloppy and sometimes crude but he can sleepwalk his way through a mining promotion. Walsh knows all about drilling procedures and assay results and resource estimates. He had written, reviewed and signed dozens of technically worded press releases, well before Busang was ever invented. As founder, president and chief executive officer of Bre-X Minerals, he persuaded an awful lot of people to buy his stock. He knows how to hype a company and where to look for money. As far back as the mid-1980s, says Keith Miles, Walsh's old drinking buddy, he'd "call and say geez, you should get some stock, it's going to be good." He'd "talk rapid-fire, quoting numbers and holes and things."

Defensive and tough-minded, Walsh would blast anyone who ever cast the slightest doubt on his greatest treasure. After the first independent tests at Busang turned up nothing but dirt, Walsh tried his best to persuade investors not to dump their Bre-X shares, saying that he was "one hundred per cent certain" the gold was actually there. "When the first ounce of gold is poured at Busang, I'm sure the naysayers will complain about the colour," he said. His company was the victim of a "well-oiled disinformation campaign," that "big money, smart money" was in the market buying Bre-X shares.

And yet, when the awful day came in May 1997, Walsh turned around and denied any responsibility for the fraud. What did he know about mining? "I was only the money man," Walsh said again and again, as if repeating it often enough would get him off the hook. *Fortune* magazine called him the "Mr. Magoo of mining." It would take a moron, the magazine claimed, to be in Walsh's position and not know what was going on.

Walsh just shrugged. Empty head, white heart. That was his defence. "Four and a half years of work, and the pot at the end of the rainbow is a bucket of slop," he said. And $40 million. That was the fortune he and his wife cleared trading Bre-X shares from late 1995 to the end of 1996, according to reports filed with the Alberta Securities

Commission. Walsh says it wasn't nearly that much but the numbers are there, for everyone to see. He says the money he made was "chickenshit" compared to what he could have realized if Busang had been the real deal.

Deny, deny, deny. Spend, spend, spend. After the scam was exposed, Walsh drained the Bre-X corporate account, dropping $8 million on lawyers, private investigators, salaries and other expenses. He won court protection from creditors, retained a five-person staff and continued to pay their wages. He gave himself and his staff a performance bonus. He ran up a $59,000 debt using a Bre-X credit card. Some shareholders were appalled; it looked like Walsh was busy building a legal defence, using their money. A $1.2-million private investigation, which Bre-X paid for and tried to keep from the public, blamed the fraud on the company's exploration manager, Michael de Guzman, and four of his close colleagues.

This, Walsh insisted, was the brain trust that tricked him, his family, his friends, his employees, his technical advisors and every investment analyst, banker, Indonesian heavy and mining magnate who cast covetous eyes on Busang. Four Filipino men, all of them missing or presumed dead. No one else had anything to do with it. End of story.

A tidy explanation, and quite unsatisfying. For Walsh, however, it would have to do. By October 1997, he was back drinking at his old Calgary haunts. He was seen driving around in an expensive car. Bre-X had slid into bankruptcy, but its sister company, Bresea Resources Ltd., was still flush. Walsh said he was thinking of taking the $20 million or so left in the Bresea account and buying some oil and gas wells. He might write a book, he said, to counter all the "lies" written about him. But Walsh has always had trouble separating fact from fiction.

By all accounts, David Gordon Walsh was an utterly unremarkable kid. Born in 1945, he grew up with three sisters and one brother on the scrubby side of Westmount, an English enclave in Montreal known more for its large, stone mansions than the Grosvenor Street row house where the Walsh family lived. David's father, Vaughan, was a stockbroker, just like his grandfather, but he didn't make much money. According to an old family friend, the Walsh house was "always messy. Mrs. Walsh was not the most diligent housekeeper," says Sharyn Scott. "But we partied a lot at David's house. We drank a lot. Everyone did."

Walsh was an outcast at Westmount High School, and not much of a student. It was the early 60s, and he was a smoker, into motorcycles and leather jackets, the Rebel Without a Cause look. His nickname, Scott says, was Iggy. After high school, he went straight into the investment business, managing individual estates, stock and bond portfolios for the Eastern Trust Company. It wasn't long before he met a fiery young secretary one year his senior named Jeannette Toukhmanian. Jeannette, known as Jaja to her friends, was a little snip of a woman, barely five feet tall, and less than one hundred pounds. Born in Alexandria, Egypt, she had come to Canada as a child with her Armenian parents, Erwant and Shake, and her brother, Robert. Jeannette had a spark, a zest for life. Her ambition rubbed off on other people; she was the best thing that ever happened to young David Walsh.

They were married in September 1968 and moved into the Walshes' basement on Grosvenor. David put his head down and worked; his job wasn't glamorous but he took good care of his clients. His bread-and-butter customers, the so-called Westmount Rhodesians, were nervously guarding their wealth as Quebec's French-speaking majority began to grasp the levers of political and economic power. Those Anglos who hadn't fled the province were interested in security above all. They wanted simple portfolios that offered safe and solid returns on their investment. Walsh managed to satisfy them and in 1969, he was named head of Eastern Trust's investment department. One year later, the Walshes' first son was born. They named him Sean. In 1972, they had their second boy, Brett. Their family was complete. Two years later, Walsh bought his childhood home from his parents for $47,000.

His first brush with trouble came in 1976, after his company was taken over by Canada Permanent Trust Co. One of the personal accounts Walsh managed belonged to a crooked stockbroker from Yorkton Securities. The broker was stealing securities from his clients and dropping them into his account at Canada Permanent Trust. When the broker was caught, people wondered why Walsh hadn't sounded the alarm. Three months later, he quit the firm. In August, Walsh landed a better job with Midland Doherty Ltd. Instead of managing small portfolios, he was selling large blocks of stocks and bonds to institutions. Pretty soon he became one of the top producers in the office and was given the title of vice-president. "I was the youngest individual ever made a V.P.," Walsh boasted.

But there were problems on the home front. In July 1978, Jeannette filed for divorce, alleging mental cruelty and adultery. "Like a lot of men, David was thinking with his little head, not his big one," says Sharyn Scott, admitting that she told Jeannette about David's affair. "It's the last time I'll ever do that," she adds. "It devastated her." Along with custody of their two sons, Jeannette requested weekly payments of $425, in addition to a lump sum of $25,000. The divorce never proceeded, however; David and Jeannette decided to keep their family together. They sold their house and moved into an apartment down the street.

By 1980, Walsh says, he was pulling in more than $100,000 a year in commissions. One year, he almost hit $200,000. But the routine was getting to him. The amount of money he handled kept growing, but the job was essentially the same, day in, day out. The real challenge, he reckoned, and the real money, belonged to the guys who ran their own companies, not to the traders who flogged stock on the market.

Walsh thought about starting up his own oil company out west. He'd call it Bresea, after his two sons. He didn't have any real strategy, other than to go bargain hunting for a few small oil and gas properties out west, cast-offs that the large petroleum companies couldn't be bothered with any more. Those were great days for Big Oil. Thanks to an international cartel, world petroleum prices had zoomed from $2.50 a barrel in the early 1970s to $40. With prices that high, developing Alberta's tar sands and building enormous drilling rigs in the Arctic Ocean suddenly became feasible and shot to the top of the majors' agenda. That left lots of crumbs for the little guys. Walsh figured he could find a couple of nice producing wells and watch the money pour in buckets.

The more he thought about it, the more he itched to move west. Walsh kept hearing how Calgary was booming. People there bragged about the city's phenomenal expansion. Entrepreneurs, lawyers, geologists, engineers, construction workers, cooks, strippers, you name it, people of every persuasion were pouring into the city. A moron could find a job, quit and get hired somewhere else the next day. The city's population jumped by twenty-five hundred people a month. Oil companies and banks battled to out-build each other. The skyline was dotted with cranes. There was talk of oil hitting $100 a barrel. Investors were cleaning up, at least in the short term. The market was so good you could pretty well throw darts at the stock listings and make money. Montreal, on the other hand, was gripped by the bitter and confusing politics of language and sovereignty. There was an air of resentment about the place. The city was about to enter a long period of decline, and everyone in the investment community knew it.

It took Walsh two years to prepare for the move west. On 30 October, 1980, he incorporated Bresea in Alberta. A year later, he privately sold 320,000 shares in the company to seven friends from Montreal, at a dollar apiece, and then got a listing on the Vancouver Stock Exchange. He bagged an $80,000 personal loan from the Royal Bank of Canada, snapped up a bunch of shares for himself and bought a small interest in ten oil and gas properties in King-fisher County, Oklahoma. Then he convinced his bosses at Midland Doherty to let him set up a new institutional salesroom in the company's Calgary office. Walsh figured he'd spend a few years getting Midland's salesroom going and make a nice pile. In the meantime, he'd grow Bresea into a prosperous little oil company. It didn't work out that way.

Calgary has always been an intensely entrepreneurial city, owing, some say, to its substantial American population and its resource-based economy. But there's more to it than that. Calgarians have an innate boosterish spirit, born out of isolation and nourished by competition with other major western Canadian centres such as Winnipeg, Edmonton and Vancouver. Growth has always been valued above all else. Unfettered capitalism is king. It's a promoter's paradise, built on hope and hyperbole.

The city had already experienced one oil boom. Apparently, no one learned anything from that brief period of speculative insanity. In 1913, a syndicate of local businessmen claimed to have discovered oil about fifty kilometres southwest of Calgary, which was then just a motley little railway town of ten thousand opportunistic souls. News of the strike spread immediately. Twenty new oil companies were incorporated in Calgary within the next week. "Knots of men collected on the street corners and talked petroleum," noted the *Calgary Albertan*. "John D. Rockefeller certainly would have felt himself in a sympathetic atmosphere could he have been here to breathe the oil-impregnated air of Calgary."

Except there was no oil. The original syndicate had stumbled across a pocket of naphtha, a flammable liquid with properties similar to gasoline. No matter. People crowded the sidewalks, clamouring for gilded share certificates. Vending stands were quickly installed in hotel lobbies, drugstores and barber shops. Practically everyone got in on the frenzy. Half a million dollars changed hands every twenty-four hours. According to one report, a local police officer was "retaining only one set of shoes, one suit of underwear, and one suit of clothes, having pawned all the rest of his personal effects to invest in leases and oil stocks."

By 1914, there were more than five hundred oil companies in Calgary. The vast majority were straight promotions that ended in failure. The most spectacular was a fraud perpetrated by one George Edward Buck, an Ontario-bred insurance salesman turned hell-and-brimstone preacher. Buck founded Black Diamond Oil Fields Ltd. and began flogging shares from the pulpit of his church. Within a few months, his congregation was tapped out, his company was broke, and the drill crew was demanding payment. Buck was desperate and resorted to the oldest trick in the book. Later, people would express amazement at how easily they had fallen for the scam.

One night in April 1914, Buck instructed three employees to pour a bucket of crude oil into the Black Diamond well. The next day, he invited a couple of reporters out to the site, hauled up the bailer, and, to everyone's amazement, oil spilled all over the ground. The next day the newspapers trumpeted the "Black Diamond strike," and the feeding frenzy began.

Months passed before Buck was caught out. He played the media like a fiddle, always hinting at more big news. Only a few selected guests ever made it to the site. "The well was surrounded by a ditch full of water and the river," Buck's rig captain recalled later. "Only after Buck gave the password was the drawbridge let down and the visitor given safe conduct across the moat.... It was impossible to see what was going on inside." A menacing guard armed with a rifle would fire "four or five rounds when visitors approached from the opposite bank. This gave an air of mystery to the whole outfit. Buck encouraged this activity."

The scheme was finally exposed about the same time as Calgary's anti-climactic oil boom came to an end. A handful of companies survived and eventually struck oil, building the foundation for a lucrative industry, but Buck had nothing to do with it. He abandoned his family, fled to Kansas and started a new oil company. In 1917, he was extradited back to Canada and was convicted of fraud. The ruling was thrown out on appeal. So ended Calgary's first salting scam. It would not be the last.

Ted Carter has spent a lot of time around small-time Calgary stock promoters and figures he's a "pretty good judge of character." An experienced trader, he writes a popular daily stockmarket newsletter out of Calgary called *Carter's Choice*. He clearly remembers the first time he encountered David Walsh. It was in 1984, at the Three

Greenhorns, a downtown bar favoured by the city's low-ball brokers and promoters. "It was after the regular lunch hour," says Carter, "and I walk in and there's this guy sitting in the corner, nursing a drink and looking a little glum. I walk up to him, there's no one else in the place, and I say, 'Hey, you're looking a little glum.' And we start talking."

The oil boom had cooled off considerably by the time Walsh and his family arrived in Calgary in 1982. Foreign oil producers had opened their taps again, flooding the market with cheap, light crude. Domestic oil prices fell by more than fifty per cent. Canada's federal government introduced its National Energy Program, which forced Alberta's oil producers to sell their product inside Canada at below-market rates. The net effect was a massive shakedown in the industry. Unemployment surged. Investment dried up. The money was simply gone.

For his part, Walsh couldn't adapt to the business culture inside Midland's Calgary office. "It was quite different than in Montreal," he told me, back when Bre-X was still Bay Street's favourite mining company. "The institutional business was spread out through a retail office of thirty people. And I was expected to bring them all under an organized umbrella." Walsh's job required him to take a piece of every deal negotiated by the salesmen working under him, something he says bothered him. "I've never put my hand in someone else's pocket. It wasn't my style to try and take business that was already going to salesmen in the office," he says. "So I went out and cold-called. I had also kept some of my Montreal business. But 1982 was not a good year on the market. In the fall of '82 I decided to go off on my own."

That's not the whole story. Walsh left Midland because of an internal dispute. His bosses weren't happy with his performance and, by the end of the summer, informed him they were chopping his $50,000 base salary in half. Walsh was furious. He shot back with allegations that they had reneged on their promise to staff-up with research analysts and economists. Take the pay cut or get out, Midland replied. On September 30, Walsh stormed out of the

Midland office. Two months later, he sued the company for wrong-ful dismissal.

Suddenly, Walsh found himself out of a job, embroiled in a costly lawsuit, stuck in a city that was experiencing the worst economic hangover in its history. All he had was Bresea, with its lonely little Oklahoma wells. When Carter ran into him, Walsh was struggling to make ends meet. His only income was from "consulting fees" drawn from Bresea's cash flow; a year earlier, these amounted to just $10,000. The Walshes were depending on Jeannette's income as a secretary to get by. For reasons that Carter says he never really understood, Walsh was blaming the VSE for his problems. "He was going on about the Vancouver Stock Exchange, crying the blues about how the people over there were giving him a hard time with this little company of his," Carter recalls. "It was your typical pro-moter's tale of woe. But I found myself getting kind of involved. I mean, David had a way about him. Likeable, I guess. The kind of guy you find yourself rooting for."

The petroleum industry was in the dumps, and there was little speculative money left in Calgary. Still, Walsh stuck it out, buying a small interest in seven more wells, these ones in Claiborne Parish, Louisiana. "The last thing he wanted to do was go to work for some-one else again," says Carter. "I'll say this for him, he hung in there when other guys would have given up the ghost. That is one thing you must remember. A promoter will go through hell to finally get to where he is going."

Walsh had built a close network of friends to help him through the bad times. They included Carter and Keith Miles, a geologist from Saskatchewan. Miles used to hang out with Walsh at the Three Greenhorns. "I used to see Walsh there quite often. We'd meet for beers after work. We were probably all promoting our stock or doing deals. It seems we were always out there raising money."

Both Miles and Carter remember lending Walsh a hand to drum up investor interest during an impossibly bearish stretch in the market. "Everybody wants to create activity in his stock," explains Miles. "David would call and say he needed a little help. I'd do the same thing. You'd get a few guys to buy the stock. You'd get some good results and send the reports around, and people would buy the stock. That's how the deal is done. It's done all the time. It's not insider trading or anything," Miles says. "Just the guys helping out."

Walsh buttonholed Carter once at the Three Greenhorns. "I walked in and Walsh was sitting there, talking to a friend of his. He turns to me and says, 'Hey, here's a guy who will buy something.' He told me he needed to build up his shareholder list. Well, just sitting at the bar that day I think I got him another eight subscribers." If he was going to attract any serious investors, however, Walsh had to get new properties. The trouble was, Bresea didn't have any cash. The royalties that trickled in from its oil wells went straight to the Bank of Montreal, holders of a US$124,000 production loan. Bresea went virtually dormant; in 1985, it made just one deal, shelling out $5,000 for a well in southern Alberta. By then, Walsh realized that he'd never be an oil-baron.

"If you ask me, David Walsh is a prick." Barry Tannock heaves his considerable bulk around in his chair and grimaces. "Of course, I've been suing the son of a bitch for years, so, yeah, my opinion of him may be slightly tainted." Tannock didn't always have such harsh things to say about Walsh. The two men were once extremely close; they've known each other since the early 1980s. By the time their relationship soured, Tannock had lost all faith in his old friend. Walsh "is hard, in a business sense," says Tannock. "If he sees an opportunity, he'll take it. He's loyal selectively. I'm not sure what his criteria are."

Tannock first encountered Walsh in an upscale hotel bar called the Owl's Nest. He doesn't recollect what they talked about, only that their chance meeting was "very convivial." Tannock had recently ended a long career with Shell Canada Ltd., where he had started as a computer programmer before settling into the accounting department. Thinking he could do better on his own, he went into the "consulting" business, advising small companies on pay-rolls, computing and investor relations.

In 1984, Tannock began doing some clerical work for a junior oil company, Argus Resources Ltd. Walsh was running Bresea out of the same office, just a few blocks from the Three Greenhorns. Business was slow, and the two men had plenty of time to kill. They spent hours in the bar dreaming up new plays. Walsh and Tannock would frequently get to talking about the mining sector. "Neither of us were really doing very much," Tannock says. "David had Bresea, but it was just limping along, collecting some piddly revenues from the Louisiana wells. Oil and gas looked like a non-starter. But mining looked interesting."

Walsh and Tannock decided to join forces and form a new company, Ayrex Resources Ltd. Walsh would be chairman and in charge of raising capital. "David was always good with new issues," Tannock says. His job, as president, was to look after the books, deal with regulatory issues and prepare press releases. Most important, he would bring his cousin Stanley Hawkins to the table. "Stan had this company, Tandem Resources, which had some good properties," Tannock explains. "David and I had nothing. So we went to him."

Hawkins has three typical attributes of a penny-stock promoter: unfailing energy, boundless optimism and questionable taste. His office, on the seventh floor of Toronto's First Canadian Place, is dominated by a large fibreglass waterfall festooned with artificial plants. The get-up resembles a set piece from "The Flintstones." Hawkins' specialty is staking territory around hot mineral plays. It's an old game, designed primarily to lift a company's stock. Stand next to a winner and someone might fling a few dollars your way. Occasionally, you might actually luck onto a real deposit of your own. In Tandem's case, that hasn't happened. In the past fifteen years, the company has staked ground near some of the world's hottest exploration areas, looking for everything from gold to diamonds. It has yet to strike serious pay dirt.

In fact, Tandem loses money just about every year. Its only revenue comes from oil and gas properties owned by a subsidiary. And yet the company does more than just survive; it issues, in a good year, millions of dollars' worth of new shares, maintains a presence in the tallest office tower in Canada, and spends half a million dollars annually on assorted administrative expenses and fieldwork. And it's by no means unique; Tandem is just one of hundreds of Canadian mining companies that seem to do little more than issue press releases and trade stock.

"What we do is gamble," says Hawkins, bluntly. "You have to raise money to fund projects. To do that, you have to do a certain amount of promotion, if you want to call it that. You need a network of brokers. Some people will listen to you, some won't. You try to get them to buy your stock and raise money for it. That's exactly what everyone does. That's exactly what David learned to do."

Tandem was a very popular stock back in 1985. The company got involved in the Casa Berardi play in northeastern Quebec, which had become famous after a large gold discovery in 1983. "We had a lot of ground in the same general area," says Hawkins. "People were throwing money at us. Walsh had Ayrex, and he wanted to get involved in Casa Berardi. So he asked for my participation."

In August 1985, Hawkins granted Ayrex 132 claims in Abitibi County, Quebec, near the Casa Berardi find. In exchange, Ayrex promised Hawkins $31,398 and 250,000 shares from its treasury, once it went public. Walsh and Tannock bought another thirty claims from Hawkins for a flat $2 and sold them back to Ayrex for $400 and 750,000 shares. Early in 1986, Ayrex successfully completed a public offering of one million shares at 30 cents each and earned a listing on the Alberta Stock Exchange. On paper, Walsh and Tannock had already made money.

Walsh then set about promoting his little company. A preliminary exploration program resulted in this upbeat news bite from Ayrex in May 1986: "Indications suggest five groupings pointing to drilling targets. These groupings are based on visual and assay gold results. . . . " The actual results were kept a mystery, however. Five months later, Ayrex announced more "encouraging results." Again, no details were forthcoming. Still, it was enough

to attract a private placement worth $900,000. This allowed Walsh
to buy more property from Hawkins, this time in northwestern
Ontario. He also bought an interest in a 26-square-mile (6,734-
hectare) concession in northeastern Quebec, near an alleged platinum discovery.

With the Ayrex promotion well into play, Walsh refocused on
Bresea. The VSE was still giving him grief, this time over a mysterious forty per cent decline in the price of the company's stock.
According to media reports, officials with the Vancouver exchange
believed that the steep drop may have been engineered. Bresea
wanted to issue a new block of director and employee options.
Normally, new options are issued at the price the company's stock
last traded, which meant Walsh and his friends were looking at
scooping up a bundle of bargain-basement Bresea shares. The VSE
ordered Bresea to increase the strike price — the price at which the
options could be purchased — on the options, so that they would
"properly reflect the market." Walsh balked. He pulled Bresea off
the VSE and moved it to the Montreal Stock Exchange.

Privately, Walsh doubted Bresea was going anywhere. "Things
sorta slowly died down," he told me. "The flow-through money
dried up." He listed Bresea in Montreal in November 1987, just a
month after the biggest stock market crash in history. Walsh knew
"there was no opportunity to raise junior mining money."

No matter. Walsh relentlessly hyped Bresea's potential anyhow.
The company acquired another proximity play, an 89,500-acre
claim in the Northwest Territories that wasn't far from a modest-
sized gold strike. The new property never amounted to anything,
but Walsh tried to stay keen. According to one colleague, he became
"quite animated when he talked about it. He'd say, 'This is the thing
that's going to put the company on the map.' He talked as if he
really believed it." He even enlisted the help of his wife, recalls
Keith Miles. Jeannette was working "behind the scenes," helping
promote the company. "I heard her talk to brokers and tune them
right in."

"I am very enthusiastic regarding the outlook for Bresea," Walsh wrote in Bresea's 1987 annual report. "A major effort has been made to enhance our public profile. With a demonstrated ability to acquire highly prospective properties…the coming months will be very exciting." On the surface, Walsh appeared to be hard at work, running a promising little exploration company. Once again, the truth was a rather different story.

Walsh was spending most afternoons in the Three Greenhorns, sucking back Budweisers and cigarettes. According to Ted Carter, Walsh drank regularly and heavily, often downing five, ten, a dozen drinks in a single session. "He drank with a passion," says Carter. "But we all did in those days. There'd be these great lulls, when people were waiting for drill results, financings, what have you. We had time on our hands. So we would go out for drinks. It could stretch from noon hour to supper and beyond."

When he wasn't hanging out in the bar, Walsh was flipping stock. "It's the only way I survive when there's no salary coming to me from various companies I'm involved in," he explained. His specialty was buying shares in new companies, just as they entered their "honeymoon" phase on the market. He'd sell them a couple of weeks later, before the bloom left the rose and the shares started their inevitable descent. At no time, he says, did he have any intention of hanging on to the stock; he was an aggressive, professional trader, trying to scoop some quick bucks. Walsh had trading accounts with Pemberton Securities, Midland Doherty, Merrill Lynch, Richardson Greenshields, McDermid St. Lawrence, Wood Gundy, and at least half a dozen others. He says there was "probably not" a single brokerage with which he did not have an account. Some were in his name, some were in Jeannette's, some "belonged" to Brett and Sean, but he handled them all. According to Jeannette, her husband was trading shares in her name without telling her. "I had no idea," Jeannette admitted. "My husband had lots of accounts for me in my name [but] I didn't know I had an account in Pemberton, nor at Midland, nowhere else."

Meanwhile, the family's financial situation was growing perilous. Money was tight. The Walshes were having trouble making loan payments to two banks. Their only regular source of cash was Jeannette's secretarial income, but this amounted to less than $20,000 a year, hardly enough for themselves, let alone their two teenaged boys. By the middle of 1987, Walsh says he was virtually tapped out. He'd invested most of his market earnings in Bresea and Ayrex. His only other assets were the family's mortgaged split-level house in a middle-class Calgary suburb, and a black 1979 Buick. He would later explain that he "basically put off the two banks until it got to what I would call a crisis, where they definitely wanted me to make some capital payments." The crisis came in July, when the banks began demanding payment. "One way or another," Walsh said, "I had to get my hands on some cash."

On 6 July, 1987, Walsh made a bad judgment call. That was the day Douglas Mair, a broker with Pemberton Securities (Pemberton was later bought by RBC Dominion Securities Inc.), called to tell him that a company called Cinram Ltd. had split its shares three-for-one. Jeannette's account at Pemberton was being credited with three thousand new Cinram shares.

Walsh knew that Mair had made a mistake. Jeannette no longer owned any Cinram shares; he'd already flipped the stock for a $7,375 profit. "I said I didn't know what he was talking about," Walsh recalled. "There's no shares of anything. He says no, he's checked and checked with accounting. I said, 'Fine, send them over to the bank.' He says, 'What are you going to do with them?' I said, 'I'm going to sell them.' He said, 'I'll sell them.' So he sold them. One phone call."

One week later, Jeannette received a cheque for $50,690. She cashed it the next day, and the money was quickly put to use. The Walshes paid the Royal Bank $15,000 in back loan payments. They put another $15,000 into a line of credit. Jeannette took about $1,200 and flew to Florida to attend a cousin's wedding. When she returned to Calgary two weeks later, she quit her secretarial job at an oil company. Walsh moved Bresea and Ayrex out of an office he'd been sharing with another promoter and into a new space. The rest of the money was "spent on living expenses."

Pemberton soon realized its mistake and asked Walsh to return the money. He refused. Pemberton sued to get it back, accusing Walsh of fraud. Walsh said it wasn't his problem; Pemberton had goofed, not him. In an exchange with Pemberton's lawyer, he also testified that the $50,000 had meant nothing to him and Jeannette, despite the fact they were flat broke when the cheque arrived.

LAWYER: What was your financial situation?
WALSH: No cash. Debts. I had been in debt for several years after leaving Midland Doherty.
LAWYER: Did [the $50,000] seem like a large sum of money considering your financial situation at the time?
WALSH: No, it seemed insignificant to me.
LAWYER: You understood when this telephone call came through you were getting sort of a windfall?
WALSH: Actually, at the time my thought was 'Figures, another incompetent back office at a brokerage firm . . . and that I had some stock sitting there that shouldn't have been. I figured it was a screw-up . . . but as I said, I didn't give it too much attention.
LAWYER: Was it alarming to you, sir, the suggestion that you were going to have to repay the $50,000?
WALSH: It wasn't a very pleasant thought, that's for sure.

Walsh could not explain why he spent the money when he knew it did not belong to him. "Keep in mind, I was also running two junior resource companies," he sputtered. "You have to have your priorities, which basically weren't worrying about bank statements when you are putting out fires running a junior resource company."

Not surprisingly, the judge presiding over the civil case didn't buy it. Walsh was ordered to repay Pemberton. He never did. "It used to bother me," Walsh told me. "It doesn't bother me any more. Life goes on."

In his case, life kept going from bad to worse. The Walshes were living on "borrowed money." They began racking up heavy debt on fifteen different credit cards, waiting for the day their luck would change. Time was running out on David's promotions. Ayrex was dead in the water. Exploration drilling near Casa Berardi turned up nothing. Walsh and Tannock owed Stan Hawkins a couple of hundred thousand dollars, a debt they had no hope of repaying in cash. Instead, they gave him control of Ayrex.

Bresea was listing badly. In 1989, Walsh reorganized the company, basically turning it into a public holding company with interests in a handful of different promotions. The objective was to spread the assets around as much as possible, increasing the potential for fund-raising. Some of the schemes were bizarre. A new Bresea affiliate became involved in a controversial immigration plan that gave foreigners the chance to obtain Canadian citizenship in exchange for large investments inside Alberta. Walsh planned to broker oil and gas properties to eager "Far East" immigrants. He abandoned the strategy within months, having failed to attract any investors, despite the "billions of dollars' worth of oil and gas assets in Western Canada for sale."

With time running out, Walsh made up another new company. He called it Bre-X Minerals Ltd. Listed on the lax Alberta Stock Exchange, Bre-X had nothing to pitch but some exploration properties obtained from Bresea, using money Walsh had raised privately. He drummed up $50,000 among friends and family back in Montreal. Three of his sisters bought 100,000 shares each; so did his mother, Kathleen. The price: one penny per share.

Despite Bre-X's dubious prospects, officials at the ASE were happy to list it. Many of the new junior companies that appear on the exchange every year are "blind-pools." They have no assets at all, not even a business strategy. The ASE sees itself as a kind of incubator of capitalism. Unfortunately, it also attracts its fair share of bottom-feeding promoters out to make a fast buck manipulating stock. "We are quite familiar with the subculture of humanity that happens to be attracted to a market like ours," exchange president

Tom Cumming says. On the plus side, he says that the ASE "teaches" junior companies "how to be a public company."

Walsh was still learning how to run one. Bre-X was virtually a one-man show, and he needed help. He turned to a former journalist and newsletter writer named Christine Starnes to help promote his company. Starnes claimed to have extensive contacts in the United States. Walsh convinced her to accept Bre-X shares as payment in lieu of cash. Walsh asked a friend of his, John Thorpe, to look after the books. A retired accountant, Thorpe was reluctant at first. He'd just sold his private practice to a major accounting firm and was pursuing the quiet life, spending his summers at a mountain golf resort, enjoying life as a course marshall, chasing the odd grizzly bear off the fairway. But Walsh convinced him to come in to the Bre-X office once a month.

"It wasn't exactly onerous work," says Thorpe. "There'd be maybe one deposit a month, five cheques coming in and out. I probably billed $500 the first year. Actually, I don't think it was even that much. There were huge blocks of time with nothing to do. David spent a lot of time in the bar. I don't think there was ever much interest in the Quebec stuff, or the Northwest Territories stuff."

Walsh started playing with the notion of buying three ancient mines in a place called Oatman, Arizona. Trouble was, the mines, which had first gone into production back in 1847, had been closed for almost fifty years and were in dangerous condition. Bre-X would need US$350,000 just to rehabilitate the crumbling shafts. Walsh tried to create interest by suggesting that the mines still contained gold, as much as 300,000 ounces. Bre-X was "poised to make a rapid transition to being a gold producer," he wrote.

In fact, Bre-X was about to slam into a brick wall. The company posted losses of $692,000 between 1990 and 1991. Total revenues amounted to less than $2,000. By 1992, Bresea had an operating deficit of almost $2 million and its shares were barely trading. Officials at the Montreal Stock Exchange informed Walsh that the company would lose its listing if it did not become active. Meanwhile, Pemberton had won a judgment to get its $50,000 back. An old loan from the Royal Bank was also in arrears. Walsh was forced to move his operations out of downtown and into the basement of his house. "David was wandering in a sea of mud," says an old friend of his, "Fast" Eddie Schiller. "He was still trying to arrange the next deal, looking for rainbows. He was really hurting."

David and Jeannette began preparing for bankruptcy, citing, among other things, "low income from self-employment" and "excessive use of consumer credit." That was an understatement. By 1992, the Walshes had drummed up over $59,000 worth of credit chard charges. The debt was spread over twelve different Visa and MasterCard accounts, two from Woolworth's department store and one from Canadian Tire. Their only discernible assets were several hundred thousand shares in Bre-X and Bresea, valued at less than $15,000, some household furniture worth about $1,000, and the old family Buick, appraised at $1,000. After Walsh had taken care of legal bills, taxes and household expenses, there was less than $3,500 left to satisfy the rest of his creditors. Pemberton received $1,500. The credit card companies split the rest among themselves.

Walsh had no choice but to shuffle the corporate deck once again. Bre-X, he announced dramatically, would quit the mining business and "focus solely on the acquisition of producing oil and gas properties in Western Canada" instead. The company would change its name, "to more readily identify with the energy-related business." Bre-X stock bottomed out to 25 cents. Walsh has hanging on by his fingernails.

Then, incredibly, he got a break, a tiny, circuitous bit of luck when Diamet Minerals announced its diamond find in the Lac de Gras area of the Northwest Territories. The site just happened be a mere forty-five kilometres from Bre-X's stagnating gold claims. Walsh seized the opportunity. His next shareholders' report was jokey, aggressive, decidedly upbeat. And shameless. "Yes, we are still in business," he proclaimed, "and proceeding with a sound strategy to develop your Company into an intermediate-size natural resource company as expeditiously as possible." Diamonds. That's what Bre-X was all about now. Suddenly, it was sitting on some pretty good rock, thanks to a quick "reinterpretation" of airborne geophysical surveys that "identified what are believed to be high-quality kimberlite pipes within the claim boundaries."

It was nonsense, of course. But Kennecott Canada Inc., a subsidiary of RTZ Corp. PLC., the world's second biggest mining company, could afford to take chances. It optioned almost the entire Bre-X claim area, giving it a seventy per cent interest in the property in exchange for cash. The agreement got Bre-X's shares moving again. The monthly volume picked up from a measly 3,000 shares to 420,000, boosting the stock's average price to a quarter. It wasn't much, but enough to keep the company going a little longer. "All of a sudden David went from the gold business to the diamond business," laughs Barry Tannock. "He didn't know sweet tweet about diamonds. But that's what sustained Bre-X for the next year."

Walsh didn't waste the reprieve. He got on the telephone and started cold-calling old friends and colleagues, anyone he could think of who might be able to offer him a deal, get him started on something else. One of the names on his list was John Felderhof. Beside his name, he'd scrawled the word "Indonesia."

5

WAKING
THE GIANT

*Well, we had to use force. But this did not mean that we just
shot them, bang bang, and were finished with it. No! Those
who resisted, yes, they were shot. There was no other choice,
because they resisted. Some of the bodies were just left where they
had been shot. This was meant as shock therapy so that people
would realize that loathsome acts would meet with strong
action that was taken to stamp out all the inhuman criminal
offenses. And so these despicable crimes came to an end.*
  – Suharto, *My Thoughts, Words, and Deeds: An Autobiography*,
    page 336

*Since I believe that death is in God's hands, I feel that we,
as human beings, should not be involved in the decision of
someone's death.*
  – Suharto, *My Thoughts, Words, and Deeds: An Autobiography*,
    page 337

IT'S BEEN CALLED "ASIA'S SLEEPING GIANT,"
which is misleading, for Indonesia is a restless, frac-
tious country that boils with ethnic riots and regional
unrest, avoiding complete and utter catastrophe, but only just.
Indonesia's power-brokers work hard to keep the country's internal

affairs quiet and off the world stage. Their efforts to avert unwelcome scrutiny have helped limit serious analysis of the country's role in the Bre-X scam. Indonesia is dominated by three interests: President Suharto, the army and big business. All three conspire to run the country; all three were deeply involved in Busang. Most observers have ignored this, dismissing the Indonesian interests as minor players in the affair. Journalists reduced the country to a pathetic stereotype, a distant, exotic place overcome by cagey foreigners. The real story is more complicated, and far more intriguing.

Until recently, Indonesia was described as Southeast Asia's biggest "tiger," powered by a roaring economic engine, unfettered by annoying regulations and standards, a cheap place to do business, where a few greased palms could unlock all kinds of doors. Much of the mystery shrouding the country stems from the manner in which foreign businesses conduct themselves there. Companies that choose to exploit greedy bureaucrats, low-income workers and loose environmental laws don't broadcast their dealings back home. Shady ethics don't beget positive spin, as American shoemaker Nike recently discovered when it was revealed the company was refusing to pay its Indonesian workers a minimum wage of US$2.50 a day. The news caused an uproar in North America. In Indonesia, it passed without comment.

Investors like sunny economic indicators, and Indonesia had plenty to throw around until the late 1990s: 200 million citizens, a growing middle class, a stunning seven per cent annual growth, liberalized trade, plenty of natural resources and escalating foreign investment. But behind the impressive statistics was a corrupt financial network controlled by Suharto and his friends. Indonesia's ruling elite was running the nation's economy into the ground, borrowing money it could never repay. By mid-1997, Indonesia owed foreign lenders US$60 billion, most of which was short-term debt payable within a year. The country's state-owned banks sat on the brink of collapse, thanks to US$3 billion in bad loans handed out to political favourites. This deeply fractured nation, with its terrifying record of abuse, exploitation and mass murder, was pitching headlong towards fiscal ruin.

Indonesia is an extraordinarily difficult country to govern, even at the best of times. Maintaining cohesion while fending off powerful external influences has always been a conundrum. The archipelago spans five time zones. Its nearly fourteen thousand islands are home to more than three hundred ethnic groups who speak 583 language dialects and who worship myriad gods, despite the country's Islamic predominance. Its diversity is without parallel, unrivalled since the break-up of the Soviet Union and its satellite socialist states. Like China, Indonesia spends enormous resources trying to maintain order, both in the heartland and in outlying areas. The military's efforts to quell dissent often end up in bloodbaths.

Aceh. Irian Jaya. East Timor. The names might not elicit more than a flicker of recognition among denizens of Bay Street, but in Indonesia, these "trouble spots" have come to symbolize the cost of development, measured in human lives. All three regions have witnessed extreme levels of violence this decade, arising from local resistance to perceived foreign interests. The ruling elite see force as a necessary evil, essential to protecting the country's political and economic objectives. "We don't want to become another Yugoslavia," argues Beni Wajhu, an Indonesian businessman with close connections to the mining industry. "Yes, we have a strong army. But it is considered the guardian of our independence. Its function is to guarantee that the dreams of our founders are maintained."

Indonesia has paid a high price for progress. Most of the country was ruled by different colonial powers until 1949, when independence was finally established by a group of army generals, aided by the newly created United Nations. Since then, Indonesia has been ruled by two men, first Sukarno, and then Suharto.

A Dutch-educated nationalist, Sukarno assumed the mantle of power after the Second World War, naming himself Indonesia's "President for Life" and outlining a basic political credo called *Pancasila*, or Five Principles. It demands a belief in one supreme God, a just and civilized humanity, unity, democracy guided by the "inner wisdom of unanimity," and social justice. While *Pancasila* remains the backbone of the country's constitution, interpreting and executing its five noble principles depend entirely on the whim of the nation's one real authority.

Sukarno was a schizophrenic leader, perhaps out of necessity. He was constantly trying to appease conflicting interests. Critics viewed him either as a manipulative demagogue or as a part-time democrat. Muslims were suspicious of Sukarno's openness to other religious philosophies and feared he could be manipulated by Western interests. Sukarno and his allies rejected capitalism, associating it with colonialism, a system the entire country abhorred. Instead, he took a more socialistic approach to development, building ties to the Soviet bloc and to China, allowing the military to take command of the nation's burgeoning bureaucracy.

By the late 1950s, Sukarno was struggling to maintain an odd coalition of Muslim nationalists, communists and the army. It wouldn't hold. The nationalists despised the communists, who pressed for radical land reforms and were intent on forming their own armed units. This, in turn, threatened the military establishment. Certain factions within the army became convinced that Sukarno had turned himself over to communism. As proof, they pointed to his increasing resolve to defy American foreign policy.

In the late 1950s, the Cold War was in full swing; Washington wanted Indonesia's cooperation in its fight to contain the "Red Menace." Sukarno wasn't terribly interested. He was more focused on expanding Indonesia's own interests by grabbing control of foreign property. His first target was West Papua, the distant territory that includes the western half of New Guinea, which remained under Dutch control. In 1962, after threatening a military offensive, Sukarno managed to push the Dutch into a startling settlement, in which they would cede the territory to Indonesia, on the agreement that the people there could later vote on whether to remain part of the country. The new province was eventually named Irian Jaya, or "glorious light."

Sukarno next set about attacking Malaysian territory in northern

Borneo, and on the Malaysian peninsula, using weapons obtained
from the Soviet Union. He also expropriated hundreds of American,
British and Dutch companies doing business on Indonesian soil. In
1964, Sukarno demonstrated his anti-American attitude by telling
the United States to "go to hell with your aid." He also pulled
Indonesia from the United Nations, an unprecedented move. Early
the following year, Sukarno told *The New York Times* he had "secret
information" that indicated that the Central Intelligence Agency
was planning to assassinate him.

There was a bloody coup that year, but according to Indonesia's
official history, it never happened. Suharto, the crafty major-general
who ousted Sukarno, has always insisted that he seized power grace-
fully, in accordance with the constitution, and that his predecessor,
Sukarno, willingly and happily spent the rest of his life locked up
inside a safe house on the outskirts of Jakarta.

It's hard to find anyone in Indonesia willing to contradict
Suharto's version of events, at least on the record, because anyone
found publicly doing so is guilty of subversion and may be imprisoned, tortured and put to death. Ibnu Sutowo, the former head
of Indonesia's state-owned oil company, explained that "everyone,
without exception, does the bidding for Suharto.... He wants
everyone to follow his line one hundred per cent.... Suharto has no
interest in creative and independent actions. Look at the people
around him now. Even when they know big mistakes are being
made, they remain silent and agree. No one has any guts."

What is commonly known, but not openly discussed, is that
Suharto masterminded an army revolt in 1965, which he sold as
a "counter-revolution" to stop "insurgent" communists from grabbing control of the government. In the next twelve months, the
so-called Year of Living Dangerously, Suharto's army, along with
ordinary civilians, murdered between 250,000 and 500,000 Indonesians, whom they branded as communists.

The insurrection was barely noted in the Western media; Viet-
nam was of far greater concern. When *Time* magazine finally got
around to examining what had happened, it offered the following

gruesome account: "According to accounts brought out of Indonesia by Western diplomats and independent travellers, Communists, Red sympathizers, and their families are being massacred by the thousands. Backlands army units are reported to have executed thousands of Communists after interrogation in remote jails.... Armed with wide-bladed knives called parangs, Moslem bands crept at night into the homes of Communists, killing entire families and burying the bodies in shallow graves.... The murder campaign became so brazen in parts of rural East Java that Moslem bands placed the heads of victims on poles and paraded them through villages."

The magazine summed up the gruesome spectacle by calling it "the West's best news for years in Asia."

Was Suharto's coup backed by the American government? Some people think so, arguing that the massacre sent a strong message to other strategically placed countries that the United States considered "soft" on communism. Dean Rusk, the American secretary of state in 1965, told the U.S. ambassador in Jakarta that Suharto's army was the "only force capable of creating order in Indonesia."

Three years later, the CIA took the unusual step of publishing, for public consumption, a devastating account of the Suharto coup, which it labelled as "one of the worst mass-murders of the twentieth century, along with the Soviet purges of the 1930s, the Nazi mass-murders during the Second World War, and the Maoist bloodbath of the early 1950s." Some commentators insist the CIA's description was a smokescreen. A former CIA agent, Ralph McGehee, described it as an attempt to "conceal [the agency's] role in the massacre." In any event, there's no question that the Suharto regime encouraged closer ties with North American multinationals. Mining companies, in particular.

Suharto's long-standing favourite is New Orleans–based Free-port McMoRan Copper & Gold Inc. The company first came to Indonesia in the early 1960s, when it was known as Freeport Sulphur

Company, and not long after it had lost a large nickel mine in Cuba to the Castro government. In search of fresh opportunities, it sent a group of geologists to Irian Jaya to evaluate a copper deposit high in the snow-capped Sudirman mountains. Working with information collected earlier by Dutch explorers, the Freeport team and their hired guides spent two weeks cutting through the jungle and scaling the mountain range. As they approached the five thousand metre level, they came across a massive outcrop of rock, seventy-five metres high, bearing all the tell-tale signs of mineralization. In fact, they were looking at the biggest above-ground copper deposit in the world.

The Dutch already had a name for it: Ertsberg, meaning "ore mountain." Any thought of mining the remote site back in 1936, when it was first discovered, was superseded by the fact it was accessible only by foot. Even if that challenge could somehow be met, political tensions between the Netherlands and Indonesia destroyed any hope that Ertsberg would be developed by Dutch interests. Freeport was another story.

By 1966, Suharto was firmly in control of Indonesia and was looking for new allies to help exploit the country's natural resources. Among the policies introduced under his "New Order" regime was a foreign investment law, which reversed Sukarno's leftist development plan and gave various tax and royalty incentives to Western interests.

Thanks to some helpful brokering by American petroleum giant Texaco, Freeport's executives were introduced to some of Suharto's key economic advisors. Negotiations regarding the development of Ertsberg took several months; in April 1967, Freeport became the first foreign company to sign a deal with the new, business-friendly Indonesia, agreeing to a thirty-year mining licence in exchange for lucrative royalty payments and a ten per cent chunk of Freeport's Indonesian-based subsidiary.

The relationship has endured to this day. Freeport's mining division, which was spun off and named Freeport-McMoRan Copper & Gold, enjoys remarkably close ties with the Suharto power structure.

The company's president, Jim Bob Moffett, has better access to the Indonesian leader than any foreign leader or diplomat. They make an odd pair: Suharto, now in his late seventies, is an austere conservative, the ruthless commander of a militarized Islamic state. Moffett, meanwhile, is a fifty-nine-year-old former college football player with a bouffant hairdo, an outgoing Texan who enjoys swinging his hips to rockabilly music. Despite their personal differences, the two men have formed an indivisible alliance. There is a simple reason: They need each other.

Moffett owes everything to Suharto. Moffett's company, which reports annual revenues of US$2 billion, would not exist without the Indonesian president's blessing. Virtually all of Freeport's assets are inside Irian Jaya. The company has long-term concessions totalling 3.6 million hectares, a land mass two-thirds the size of Texas. This includes the stunning Grasberg deposit, which went into production beside the original Ertsberg discovery in 1990. Grasberg is by far the world's richest mine, with 55 million ounces of recoverable gold, 43 million ounces of copper and 118 million ounces of silver, valued at more than US$40 billion. It's a money machine, and it has made Moffett one of the top-paid chief executives in the United States. In 1996, he took home a staggering $33 million in base salary and other remuneration.

Conversely, Suharto has reason to be grateful for Freeport. The company took a serious risk investing in Indonesia during the early days of the "New Order," when other multinationals veered away. It has constantly promoted the country within the international business community, while downplaying Suharto's dismal record on human rights. Freeport's own corporate history parrots Suharto's revisionist claim that the "friendly" general reluctantly took power after communist forces tried to overthrow Sukarno. More important, perhaps, is the fact that Freeport employs more than thirteen hundred people and has become Indonesia's largest taxpayer, pumping almost US$200 million into the nation's coffers each year.

Freeport's favoured status hasn't wavered, even after a series of violent episodes involving local indigenous people drew condemnation from a cluster of high-profile human rights agencies. The worst incidents, in which members of the Amungme, Dani and Kamoro tribes have lost their lives, have usually involved members of the Indonesian military and private security forces, which protect the Freeport mine from outsiders.

About a thousand government troops patrol the Freeport operation, which extends from the mine site high in the mountains to a giant shipyard on the coastal lowlands. The company is building a new, carefully landscaped company town, complete with schools, a large recreation centre, a "retail and entertainment" complex and hundreds of freshly painted homes, which will ultimately provide shelter for thirty thousand people. It's easily the biggest commercial enterprise in Irian Jaya, and the relative wealth of the people who live and work in the area has attracted thousands of migrants. As one Freeport executive explained to me over lunch at the company's office complex in Jakarta, "Where there's honey, there's ants."

Intruders are not welcome. In 1996, the *Multinational Monitor* named Freeport one of America's ten worst corporations, thanks to its Indonesian operations. "The company operates a virtual colony in Irian Jaya," the watchdog noted. The first problems started to flare as early as 1967, when Freeport expropriated ten thousand hectares of land. The land had been used by the Amungme, who had lived and hunted in the area, undisturbed, for centuries. According to the Amungme, the mountains where Freeport dug its giant, open-pit copper and gold mines were sacred. Freeport was defiling the resting place of their ancestors with the full approval of the Indonesian government.

Freeport officials argue that most of the Amungme's concerns have been addressed. In fact, the company argues, their lives have been improved. Freeport's patronizing view of the region's indigenous people is nothing new. In 1980, a former Freeport president described the local population as "primitive" who, "unaware of such advances as the wheel and iron tools, live much the way early *homo sapiens* did a million years ago. . . . They will eat anything that walks, creeps, or crawls."

Freeport's language has been adjusted — slightly — to fit contemporary attitudes. Local people are now "resettled" with "reasonable compensation for any houses and permanent land improvements which it becomes necessary to destroy." But the company's presence in Irian Jaya continues to be the focus of protests, particularly among members of the Organisasi Papua Merderka (OPM), an Irian Jaya separatist movement. The OPM has exploited the local people's resentment towards Freeport, enlisting them in disruptive rebel campaigns, using them to stir the fires of its own discontent by marching on Freeport property and undertaking sabotage. Freeport says that the Amungme have traditionally been divided into warring camps. The company is simply caught in the middle of an ongoing tribal feud that it wants nothing to do with.

But according to the Australian Council for Overseas Aid and the local diocese of the Catholic Church, there's been a spate of beatings and killings on or around Freeport property, involving Freeport employees or soldiers whose duty was to protect the company's territory.

The worst incident in recent memory occurred on Christmas Day, 1994, following a demonstration near the Freeport mine. A local man was stabbed and shot to death by Army personnel while riding on a Freeport bus. His body was thrown into a ditch. The same day, soldiers arrested a group of men returning from a Christmas celebration and locked them up in a Freeport container. They were robbed, stripped naked and beaten with sticks and rifle butts over a period of four hours. Later, they were taken to a Freeport workshop, where three of the men were killed. All told, thirty-seven Irianese civilians were reported either dead or missing at the hands of military police during the Christmas Day massacre. Security hired by Freeport were alleged to have "engaged in acts of intimidation, extracted forced confessions, shot three civilians, disappeared [sic] five Dani villagers and arrested and tortured thirteen people," according to a report prepared by the Australian Council for Overseas Aid. Indonesia's own human rights agency, which put the death toll at sixteen, noted that five people were killed inside Freeport property.

In March 1996, *Asiaweek* magazine reported that a Freeport security officer ran over a local resident while driving a company car. Instead of helping the man to the hospital, he dumped him in a ditch and left the scene. And less than a year later, three Dani women were allegedly raped by members of the Freeport security unit, inside a Freeport apartment. This resulted in a skirmish between a crowd of Amungme and Dani men, ending in six more deaths.

And according to a Freeport official who refused to be named, another deadly incident occurred in February 1997, after some Amungme men tried to block a company road. A scuffle ensued, and "four or five people" were killed, including a Freeport employee. One man, encircled by ten stick-wielding soldiers, "freaked out and ran." He was tracked down and killed by another group of Amungme men, while the soldiers stood by and watched. "They chose not to act," says the Freeport official. "I can't say that I blame them. The situation could have grown worse."

Freeport admits there have "definitely been human rights abuses" in and around its mine, but denies any "direct" involvement. "If we leave now things will be ten times as bad," says Ed Pressman, a Freeport public relations official.

But the company can't make the same claim about the sloppy environmental practices that have ruined the Aikwa River, a key supply of fish and drinking water for native villagers. Every day, 125,000 tonnes of waste from the Freeport mine are flushed into the river, giving it a deathly grey colour. Tailings — crushed rock powder from the company's mill and concentration plant — make up most of the waste, which Freeport claims is non-toxic.

"Freeport has made a lot of mistakes in the past," said Indonesia's environment minister, Sarwano Kusumaatmadja, "but it has shown goodwill to repair the environmental damage." But before the company built a system of levies to manage the flow of tailings carried down the Aikwa, the waste spilled over the river banks, killing about 130 square kilometres of forest. The overflow problem has been reduced, but the tailings have made the Aikwa "no longer potable," according to Irian Jaya's environmental protection department.

Again, Freeport dismisses local concerns and says the tailings are, in fact, beneficial. They can be gathered and shaped into bricks, useful for building huts, and they make a fine growing medium for crops, including bananas and pineapples, says Ed Pressman. "Let's face it. Environmental groups have this deep loathing for us," says Pressman. "Their stated position is to shut down the mine. But that will never happen. The fact is, Freeport is about as transparent a company as you are going to get in this country." Journalists, however, are not permitted to visit the area without Freeport's permission and this, it turns out, is difficult to obtain. Pressman says the company accepts just one media visit a month, at best. There's a long, long waiting list.

While Freeport's strength and influence in Indonesia have never been matched, other mining companies have managed to develop large-scale projects in the country, the result of careful lobbying and deference to the Suharto government. Toronto-based Inco Ltd., the world's biggest nickel producer, was the second foreign company to sign a long-term mining contract with the regime, in 1968. The giant operation on the island of Sulawesi now produces forty-five million tonnes of nickel a year, and employs more than two thousand people.

Like Freeport, Inco makes bold assertions about its sensitivity to local needs. Although the company is considered to be a good corporate citizen, it has had its problems. Indonesian mining bureaucrat Simon Sembering went to work for Inco's local subsidiary in 1976 as an engineer. He doesn't have many fond memories. "The company was, how can I put it, rather arrogant towards its Indonesian workers," he says. "In fact, the Canadian management treated everyone badly. They did not take the time to get to know any of us."

More recently, Inco has attracted criticism for trying to remove economic migrants from certain areas in Sulawesi it wishes to mine. The plan is to uproot several hundred families and move them to a reclaimed marshland. Because the ground is inappropriate for farming, concerns have been raised that the families won't be able to sustain themselves.

Similar issues are being raised in Inco's own backyard. The company has suffered a series of setbacks since paying $4.3 billion for the giant nickel-copper-cobalt discovery at Voisey's Bay in Labrador. Relations between Inco and fifteen hundred natives who live near the site broke down in early fall 1997, after the two sides failed to agree on compensation for the land. The Innu Nation considers the area part of its historic hunting ground and has rejected a $77-million settlement offer from Inco. The Innu have called on Inco to give them royalties on any minerals they mine from the site.

Scott Hand, Inco's president, says that the mining industry is always under fire because of its unavoidable impact on people and the environment. The issues transcend national boundaries. When it comes to Indonesia, he insists that Inco has "gone beyond" the requirements listed in its mining agreement with the government. There are, however, certain cultural differences that can cause headaches. Miners, he says, "always attract whorehouses. But that doesn't fly in a Muslim country. We're always taking down the whorehouses."

Indonesia's mining industry was still in its infancy when John Felderhof landed in Jakarta in 1980. Only a handful of companies were actually in business; aside from Freeport and Inco, there were a couple of state-owned companies mining coal and tin. Identifying and developing a mineral deposit are extremely costly and time-consuming, requiring a highly trained workforce and patient investors. Indonesia lacked both ingredients. It was a situation the government desperately wanted to change. The question was how.

The country's Western-educated technocrats (sometimes referred to as the "Berkeley Mafia," owing to the large number who attended the California university) were instructed to devise a new investment strategy that would attract foreign mining interests. They came up with the Contract of Work (COW) system, which has been hailed as one of the finest pieces of mining legislation anywhere in the world. Designed to adapt to prevailing economic and investment conditions, the COW system has gone through a number of changes since it was first introduced in 1967. Each change results in a new "generation" of COW. There have been seven generations of COWs, but the basic premise has always remained the same. Foreigners are welcome to exploit Indonesia's mineral resources, as long as they share the wealth with the government. There was just one other caveat: Every COW had to be approved by Suharto. This was not always easy.

Indonesia was still unattractive to most mining companies through most of the 1970s. For one thing, there was considerable political risk. The execution of suspected communists continued across the archipelago. In 1975, the world watched in horror as Jakarta launched its bloody annexation of East Timor. A Portuguese colony five hundred kilometres off Australia's north coast, the territory was attacked by Indonesian troops and claimed as part of the republic the following year. Most of the territory's 750,000 inhabitants resisted the move. Many had seen what life was like in West Timor, the other half of the small island that Indonesia claimed from the Dutch. By Jakarta's own estimate, between fifty thousand and eighty thousand Timorese were wiped out in two years. The United Nations passed a resolution reaffirming the region's right to self-determination, but most member countries simply looked the other way. Henry Kissinger, the American secretary of state who later joined the Freeport McMoRan board, gave his tacit approval to the Indonesian offensive, suggesting that it be done "quickly, efficiently, and [without] our equipment."

The annexation didn't interrupt existing trade between Indonesia and the developed world. Foreign resource companies were more interested in Indonesia's oil and gas reserves, which were far easier to exploit than metals; a company need only drill, turn on the taps and watch the profits pour out. And a dearth of solid geological information within Indonesia turned off some mining companies on the lookout for new opportunities. After the Dutch were expelled in the 1940s, no one had bothered to map the country.

Felderhof would help change all that. He was down on his luck, coming off a difficult decade roaming the southern hemisphere, looking for mineralization in places such as Zambia, northern Australia and Papua New Guinea. Among his peers, he was regarded as something of a renegade. Felderhof had cultivated few friendships, unusual in the clannish mining industry, where everyone seems to know everyone else and fraternization is one of the profession's few easy pleasures. Like Walsh, he was a heavy drinker. According to one former colleague, "You couldn't call him at night...as he would be pissed." Felderhof's isolation came in part because of his abrasive personality, but also from the perception that self-interest, not teamwork, was his primary professional motivation.

He was a fallen boy wonder who had achieved a brief moment of fame in 1967, after he and another young colleague working for Kennecott Copper Corp. stumbled across a huge gold, silver and copper deposit in the mountains of Papua New Guinea. The Ok Tedi find was developed into one of the world's largest mines, and it now represents Papua New Guinea's biggest industrial enterprise. Felderhof never failed to mention the discovery to new acquaintances. He often complained that his role had been downplayed, that he deserved more accolades, even though Ok Tedi was not originally considered to be economically viable. "Felderhof staked his career on Ok Tedi, but that kind of thing is not uncommon at all," says one Australian-based geologist, who worked on the project for Kennecott. "It's rare to be involved in any major discovery. A lot of people spend their whole lives in mining and never find anything."

Felderhof is a blunt, impatient man, prone to sarcasm. He's not the sort to admit to any shortcomings, although he had seen his share of bad business dealings, well before he ever crossed paths with David Walsh. He was born in Holland in 1940 and immigrated to Canada with his parents twelve years later. He still speaks with a thick Dutch accent. He spent his teens in New Glasgow, Nova Scotia, a gritty coal-mining town, where his father practised medicine, and studied at Dalhousie University in Halifax, obtaining a geology degree in 1962 with a major in structural economics. He got his first job with the Iron Ore Company of Canada and was posted to northern Quebec. A year later, he left Canada for good.

After the Ok Tedi discovery, Felderhof went to work for a number of smaller mining companies in Australia. According to Tim Scott, an Australian mining veteran, Felderhof was given a rough ride Down Under. "Australia is not the most tolerant nation in the world," says Scott. "Felderhof was never really accepted. He was seen as a 'new Australian' and not one of the gang. Some of the boys ridiculed him because he was a Dutchman." Felderhof's star waned, and by 1973, he had virtually dropped out of the industry. According to some reports, he had a brief fling growing macadamia nuts.

The simple life didn't satisfy Felderhof, and in 1974 he went looking for new work, outside of Australia. He landed a consulting job with A.C.A. Howe International Ltd., a Toronto-based geological firm, run by a patrician Canadian engineer named Peter Howe. "John showed me a medal Kennecott gave him for Ok Tedi," Howe says. "It seemed to me he was a first-class geologist, so I sent him to South Africa." Based in Johannesburg, Felderhof spent the next few years working quietly on a series of projects, mainly in Zimbabwe, or Rhodesia, as it was then known. Then he got his first close whiff of a mining scandal.

In the late 1970s, A.C.A. Howe was hired by a penny-stock outfit called Leichardt Exploration to join its search for minerals along South Africa's Orange River. According to a former Leichardt director named Peter Munachen, Felderhof was directly involved in the exploration effort. In 1979, Leichardt announced it had found certain geological anomalies that suggested the presence of diamonds. The company's stock jumped from $2.50 to $15, making paper millionaires out of a few lucky shareholders. At year's end, however, it became clear there were no rocks in sight, and Leichardt's stock dropped to 12 cents. Before the Leichardt mess had been fully digested, Felderhof had left South Africa. The time had come to explore Indonesia.

Felderhof spent most of his time roaming around Kalimantan, Indonesia's huge, resource-rich territory on the island of Borneo.

Split into four provinces, the territory encompasses 550,000 square kilometres of jungle and swamp, roughly the size of France, and represents almost thirty per cent of Indonesia's total land mass. Despite its great size and wealth of natural resources, Kalimantan contains just five per cent of the Indonesian population. Even today it's one of the most remote, primitive places on earth. The Kalimantan provinces straddle the equator, in the middle of what's sometimes called the Rim of Fire, a vast string of volcanic islands off Australia's north and east coasts that include New Zealand, Fiji, the Solomon Islands, New Guinea and the Philippines.

Described in geological terms as a "tertiary volcanic corridor," the Rim of Fire was formed millions of years ago, when opposing continental plates shifted and collided, tearing land structures from the Asian continent and forcing pockets of the earth's crust to penetrate the South Pacific Ocean. Thousands of volcanic islands were formed in the process. Indonesia, a 5,200-kilometre-long archipelago, hosts the largest number of active volcanoes in the world. Although none are situated in Kalimantan, the constant heating and cooling of volcanic — igneous — rock that occurred *underneath* the region's surface, over millions of years, makes the area a likely source for gold.

In fact, gold and other precious commodities have been harvested in Borneo for centuries. According to local lore, the name Kalimantan means "rivers of gold and diamonds." There's a reason for it; diamonds are known to exist in ancient river beds, which run out from the coast and onto the ocean floor. Dutch traders built large, heavily guarded diamond warehouses on Borneo's west coast as early as the sixteenth century. In 1622, marauders from Java raided one of the warehouses and reportedly captured a 376-carat diamond the size of an orange. A major gold discovery in the eighteenth century brought some 200,000 Chinese miners to Borneo. Annual gold production for the entire region peaked at 100,000 ounces a year — about what a single small to mid-sized mine in Canada produces today — before the Chinese were driven out by Dutch imperialists.

War, drought and other distractions hampered the Borneo gold trade until well into the twentieth century, but the Dayak continued to dredge small quantities of smooth, rounded gold grains from the rivers. Their methods have become increasingly sophisticated over the years, from simple panning to the use of diesel-powered water pumps. Some of the Dayak miners were organized and financed by foreign traders, a situation that didn't sit well with the Indonesian government. By the 1980s, the army had "removed" twenty thousand "unregistered" miners from Kalimantan.

Felderhof helped kick off the second great Kalimantan "gold rush." It was more of a stock market bonanza than anything. Although a record number of exploration permits were issued in Indonesia in the 1980s, the vast majority to small Australian outfits, barely a dozen discoveries were made, and only a handful were ever developed into actual producing mines. "To be quite honest, I don't think a lot of the foreign-owned companies gave a damn if there was anything in the ground," says Michael Everett, an Australian engineer who took part in the frenzy. "As long as their shares were going up every day, they were happy."

Most companies had little trouble promoting their Indonesian properties. On paper, it made sense. The international gold price was relatively high, while exploration costs in Indonesia, with its endless source of cheap labour and materials, were exceedingly low compared to the West. And despite its wet, tropical climate, Kalimantan seemed relatively accessible, thanks to logging roads and myriad winding rivers. There was another, intangible factor. Borneo has a romantic, exotic aura about it, which Westerners have always found irresistible. Promoters exploited this fascination for all it was worth. Kalimantan, they said, was the mining industry's final frontier.

To some degree, they were right. Less than five per cent of Borneo had been mapped; the rest was *terra incognita.* By mid-decade there were more than ninety Australian companies in Indonesia, all clamouring for exploration and development permits. The Indonesian government, eager to develop Kalimantan, welcomed the Australians — their closest neighbours with any long-standing mining expertise

— with open arms. Hundreds of Aussie and Filipino geologists were allowed into the country with work permits.

Peter Howe created yet another new outfit, Jason Mining Ltd., and began raising seed money in Australia. Felderhof, his field marshal, moved slowly across Kalimantan, on foot, via river canoe, by any means necessary, talking to the local Dayak and watching where they panned for gold. "I moved west to east, all the way to East Kalimantan, building up my own geological data base," Felderhof told me in January 1997. "And during that time, you know, we identified a number of deposits." His fieldwork was very good; nobody in the industry denies it. Exploring with a trio of Australian geologists, he found at least three gold-bearing properties in Kalimantan — Ampalit, Mt. Muro and Mirah. He was at the centre of the Indo-Australian exploration boom.

As a director at Jason, endowed with plenty of share options, Felderhof had an extra incentive to help boost the company's public profile. It's since been pointed out in the Canadian press that Felderhof was not averse to exaggerating a deposit's potential, but that should come as no surprise. The mining industry is full of hype; geologists are always positive about their work, because they want it to continue. The key point is that Felderhof learned the fine craft of promotion while working at Jason. He was no longer just a wonky geologist or, as Howe puts it, "a guy who liked to tramp in the jungle." He was on his way to becoming a real player.

In 1986, Felderhof met a chubby, friendly junior geologist fresh from the mountainous gold-bearing regions of the Philippines. There were plenty of Filipino geologists in Indonesia, but Felderhof saw something special in Michael de Guzman. He was smart and cocky; he liked to brag that his IQ was over 150. Like Felderhof, he had a lot of unusual ideas about gold. Born in 1956, the fifth of twelve children, de Guzman grew up in Quezon City, a suburb of Manila. He was fascinated by rocks as a child, collecting them from local rivers and placing them on a shelf in his bedroom. De Guzman's father, a survey engineer, placed a high value on education; like all his brothers and sisters, Michael attended university, picking up a geology degree in 1979. He wed his sweetheart, Teresa Cruz, without telling his parents, the first of four secret marriages he would arrange for himself. Michael and Teresa spent the next six years in Baguio, in the northern Philippines, where he worked for Benguet Corp., an American mining company, making roughly $10 a week. That was hardly enough to raise a family, so the thirty-year-old joined the rush to Indonesia, leaving his wife and three young children behind.

De Guzman went to work for an Australian company called Pelsart Resources NL, led by Kevin Parry, a colourful businessman with an all-consuming passion for yachting. According to Michael Everett, who served time on the Pelsart board, the company fumbled around in the jungle, employing the wrong drilling techniques in their attempt to delineate gold reserves. "They achieved very poor recovery rates," says Everett. "The original drilling that was done actually undervalued the deposits."

Instead of focusing on performance, Parry's crowd was living the high life. Spending was out of control, says Tim Scott, who was hired away from a Canadian oil and gas company to work for Pelsart, at twice the salary. "I bitterly regretted the decision," he says. "I realized that Parry really didn't have the patience to develop these properties. He was throwing money away on frivolous stuff. It was all drunkenness and craziness. Typical 1980s excess." Indeed, Parry was mounting a furious campaign to compete in the illustrious America's Cup yachting race, a venture that would cost him millions of dollars. It has been suggested, but never proven, that Parry was moving funds from Pelsart to finance his bid for sailing supremacy.

Felderhof was introduced to de Guzman when Jason and Pelsart joined forces on a number of exploration efforts. From the beginning, they had a kind of professor-pupil relationship. As the older, more experienced geologist, Felderhof was impressed with de Guzman's elaborate theories. They would spend hours comparing notes and discussing arcane geology. Their favourite discussion focused on a complex mineral-bearing structure called a diatreme.

It's a type of natural depression, formed when molten lava bubbles up close to the earth's surface and is cooled by running groundwater. This ground contracts, creating a shallow hollow. Over time, the depression fills with material such as dirt, clay and plants, which eventually become mineralized thanks to constant heating and cooling. The process can repeat itself many times over millions of years, until the area resembles a mound, or a dome. If the right conditions are met, it's a slow-cooking recipe for gold, plentiful and easy to recover. Or so the theory goes.

Felderhof admits that not everyone shares his view of diatremes. He once told me that "a lot of it was perceptual thinking." De Guzman, he added, "had already developed similar ideas in the Philippines. Michael agreed with me that these structures made sense. So I hired him on as my exploration manager."

The pair figured that due to Borneo's central position in the Rim of Fire, the region must contain these complex diatreme systems. Felderhof felt there was a long string of diatremes running through the heart of Kalimantan. He pointed to Mirah and, two hundred kilometres to the northeast, Mt. Muro, as evidence. Both discoveries were classified as epithermal deposits — where the gold was found in volcanic rock — containing at least three million ounces. Felderhof's discovery of another deposit — the Kelian gold mine — 150 kilometres to the northeast of Mt. Muro gave his theory even more punch. Viewed overhead on a map, the three deposits appear evenly spaced. If one were to connect the dots with a pencil, they form a remarkably straight line. And it points directly to Busang. In April 1986, Felderhof and de Guzman made their way up the rivers and landed at Long Tesak.

The area around Busang looked much the same then as it does today. An old logging road cut a thin swathe through the jungle, north from Long Tesak to the rolling hills where Felderhof and de Guzman made camp. They were attracted by the tell-tale signs of old volcanic activity; surface extrusions indicated where the earth's

crust had been pushed upward. They paid special attention to out-
crops of bedrock along stream bed, since these exposed formations
provided a brief history of the area's mineralization and were good
indications of what lay beneath the surface. The two geologists took
a few surface samples and walked around the rivers, searching for
alluvial gold. The Dayak told them where to look.

Felderhof liked the site's potential and passed along a positive
report to Jason. But the company never followed up. Within
months, another outfit, Westralian Resource Projects Ltd., sent an
Indonesian geologist named Jonathan Nassey to have a look. He
wrote another encouraging report, noting that local Dayak were
panning for alluvial gold in the area. Westralian quickly filed claim
on the territory, along with two Indonesian partners. The new part-
nership was called PT Westralian Atan Minerals, or PT WAM. In
1987, PT WAM obtained a fourth-generation Contract of Work, giving it a long-term title to the site, providing that the company's two
Indonesian partners achieved majority control within ten years.

The first serious exploratory work started later that year. Two
Australian geologists, John Levings and Graeme Chuck, built a
small base camp in what was later known as the Busang Central
Zone and began covering the area on foot. They liked what they saw,
basically confirming Felderhof's suspicion that the area might contain a gold deposit similar to the one at the Kelian site. Another
Australian exploration outfit, Montague Gold NL, based in Perth,
decided to become involved and helped fund the exploration, which
was stepped up to include preliminary drilling.

Nineteen holes were sunk into the property. The average depth
was seventy-eight metres, which isn't much, but enough to get some
idea of what's below. Seventeen of the holes showed gold in short
intersections of about three metres or less. The average amount of
gold per tonne, described as the "grade," was calculated to be from
one to four grams. This is an excellent grade, if it comes in consistently over the length of a drill hole. The average grade at Freeport's
Grasberg deposit is a little over one gram of gold per tonne. But this
wasn't the case at Busang. The grades were completely inconsistent,

and the geologists couldn't determine whether there was really a viable deposit in the ground. They would need to do more work.

Unfortunately, the money was running out. The exploration boom was coming to a quick end, thanks to the global stock market crash in the fall of 1987, as well as a decision among major petroleum multinationals such as Penzoil and British Petroleum to divest from mining exploration and concentrate on oil and gas. Montague and Westralian invited a number of large mining companies to buy a piece of Busang, but they all passed. Speculative investment in high-risk exploration ventures simply vanished.

For the small Australian exploration companies, it meant the end of a brilliant market run and the beginning of some very tough times. The Busang project was put on the shelf. Jason sold its stake in all its Indonesian properties and eventually changed its name. Pelsart was stripped down and rebuilt. Teetering on the edge of personal bankruptcy, Kevin Parry was removed from the board, and his company was eventually passed along to Indonesian interests.

Like dozens of other geologists, Felderhof was cast adrift. He ended up in Perth, where his second wife, Ingrid, was preparing a doomed run for a senate seat on behalf of a right-wing political party. Felderhof never took advantage of his Jason shares; he left Indonesia with little more than a hacking cough. His new student, de Guzman, was briefly retained by Pelsart as a consulting geologist, but the relationship quickly soured after he was caught stealing cash and furniture from a company office. Apparently, he planned to give the booty to a girlfriend. The fact that he was cheating on his wife was beside the point. "That kind of thing goes on all the time," says Mike Everett. "But cheating on one's employer, well, that just isn't on. It was just a small amount, petty cash, really, but it was enough." De Guzman was fired on the spot. His dishonesty was a tiny harbinger of things to come.

6

HARDSELL

I want to put some romance into both companies ASAP.
*We have got a great speculative stock market going in
Canada now.*
— David Walsh, writing to John Felderhof, April 1993

*We are confident that Busang is in the bag so to speak.
Now it is just a matter to determine how big it is.*
— John Felderhof, wiring to David Walsh, April 1994

ALEX TADICH is a former journalist turned investment advisor. He has a very cynical view of stock
promotion, although he would call himself a
realist. It's no coincidence that Tadich lives in Calgary, a modern
stomping ground for some of the most notorious penny-stock touts
in Canada. In 1992, Tadich wrote a "fictional" account of the trade,
based on his own observations. *Rampaging Bull: Outfox Promoters at
Their Own Game on Any Penny Stock* is a very helpful publication,
must reading for neophyte investors.

Essentially, there are two kinds of promoters, Tadich writes. The
Good Promoter identifies solid investment opportunities, hires people and becomes a successful entrepreneur. The Bad Promoter, a.k.a.
the Rampaging Bull, "promises things that don't exist." He "always

leaves out important information in every conversation, press release, and prospectus." The Bad Promoter is a cunning narcissist who cultivates an image as a "man of the people." He will "hint at 'secret knowledge' or 'influential friends.'" A promoter's story "must be believable but not perfect." He recruits important people "by offering them options, warrants, and free trading shares." The Bad Promoter, Tadich concludes, "is a danger and menace to society."

David Walsh says he has "two philosophies in life, OPM and OPB — other people's money, and other people's brains." In the spring of 1993, he had run out of both. Few people wanted to do business with him any more. Brokers weren't returning his telephone calls. "I was so mad at David," recalls Barbara Horn, a Calgary-based broker at Nesbitt Burns. "The oil and gas stuff, the Northwest Territories thing, they went nowhere. I don't think David ever did his homework. You could never get information out of him. Nothing he touched seemed to work, okay?"

It seemed he'd used up all his lives. The new gambit up north was going nowhere. Bre-X eventually sent out a garbled press release admitting that "the diamond play weakness gives promise of being real confusing." Walsh had to reorganize. If he was going to get back in business, he'd have to come up with a new vehicle, something really unique, and promote it to the maximum. "I decided that if I was going to stay in the mining business I would attempt to hook up or associate myself with an individual I thought was tops in exploration," Walsh says. In 1993, he went looking for John Felderhof.

The two men had met a decade earlier, when Walsh and his friend, a Calgary businessman named Mike Duggan, made a trip to Australia. "Mike knew some people in the mining business and I decided to go with him," Walsh recalled. "It was an opportunity to be introduced to some people, see if there was anything of potential for us."

He can't forget that first, casual encounter with John Felderhof: "We met in an office in Sydney. He was talking a client up. We were talking in the office and he mentioned that he was off and would I like to have a look at Indonesia with him. I took the opportunity to spend, I think it was ten or twelve days in the West Kalimantan jungles."

Walsh didn't do any deals Down Under. He didn't have any money. Back in Calgary, he floated a gimmick over the news wires, announcing that Bresea had hooked up with Felderhof's boss, "world-wide mineral and geological consultant" Peter Howe, in an attempt to search for Borneo gold. Howe denies that any agreement ever existed. In any event, the market just yawned. Ten years later, Walsh decided to try again.

He'd kept in loose touch with Felderhof since their first encounter, sending the odd Christmas greeting. Felderhof passed through Calgary once during a quick trip through Canada, and they had dinner. Both men had been through the wringer and were struggling to get by. Felderhof was flitting back and forth between Perth and Jakarta, doing a bit of consulting, but, as he describes it, there was really "nothing happening." It was frustrating, because the ground had been prepared for another wave of exploration. Five solid deposits had been identified in Kalimantan the previous decade. A considerable data bank was left behind, hinting at more potential. Suharto had signed more than a hundred Contracts of Work. The land was still there. All Felderhof needed was a fresh source of money.

Walsh picked up the phone in March 1993. "I tracked John down in Jakarta. It took me a week or so. He was just winding up whatever it was he was doing, so the timing was good. I thought that nickel, sorry, gold and copper, that commodity prices had probably bottomed out, and he agreed, and I asked what was going on in Indonesia and he said nothing, that people are still pulling out. I asked if there were any opportunities and he said he thought there were. [He would need] a couple of weeks to put together some meetings for me."

Felderhof didn't refer to Busang by name, but it was at the top of his mind. The site was definitely available. Westralian and

Montague, the two companies that owned eighty per cent of the Busang COW, had fallen under the control of a Scottish entrepreneur named William McLucas. Although McLucas had helped fund the exploration boom in the 1980s, Indonesia was over for him. There were land taxes owing on the idle Busang site, and McLucas was anxious to unload the property. A dozen majors had been over the site, but there were no takers. Early in 1992, he contacted Howe, who recommended he have a pair of geologists re-evaluate the property, give it some spit and some polish. He had the perfect guys. John Felderhof and Mike de Guzman.

It was the rainy season. Kalimantan always seems wet, but in November, it pours. De Guzman spent four days squelching around Busang, sleeping in an old wooden hut left behind by the Aussies. He collected a handful of surface samples and returned to Jakarta. With Felderhof's help, he filed a report, dated December 1992. De Guzman wrote that Busang was a prime candidate for further exploration. The chance that it contained a mineable deposit of one million ounces, was, in his view, a solid forty per cent. He compared it to a large working mine in the Philippines, where he had once worked. He said that he found not one, but three diatreme systems, those elusive gold-bearing structures he and Felderhof liked to talk about. De Guzman added a little extra sparkle, describing "fine grains of visible gold" in one of his samples. Finally, he recommended McLucas spend a modest US$50,000 on a new exploration program. If nothing else, it would "upgrade the viability" of the deposit.

Busang was not a scam, not yet. "It was basically a good exploration target," Felderhof says. "Nothing more, nothing less." But de Guzman was clearly exaggerating the property's potential. No one else who had visited the site ever mustered the same level of enthusiasm. The geologists who had previously been over Busang noted that any mineralization that did exist appeared spotty. The Australians had examined fifteen hundred metres of core and couldn't come up with any reliable resource estimate of what lay below. And yet de Guzman had expressed confidence there was a viable deposit there, based on a few handfuls of dirt.

"They couldn't see the signals," Felderhof told me. "Tropical geology can be very deceptive, eh?"

Walsh needed little convincing. He took the last $10,000 from the Bre-X kitty and flew down to Jakarta with his eldest son, Sean. He also brought along a Canadian geologist named Kevin Waddell, who, Walsh wanted everyone to know, would help with due diligence and inspect the properties Felderhof had lined up. Waddell never made it to Kalimantan. For some reason, he stayed in Jakarta, reading geological reports and taking in the sights. Walsh and Felderhof talked privately, and a few days later the Bre-X advance team returned to Calgary with a deal. Walsh agreed to purchase an option on McLucas's interest in the Busang COW for US$80,000. He wouldn't wait for any "viability upgrade." He had enough raw material to kick-start a promotion. Felderhof was hired as Bre-X's general manager in charge of exploration; de Guzman would play his chief assistant. "We basically had two roles to play," Walsh said. "Mine was to finance the operation, and [theirs] was to find the gold. Pretty simplistic."

Pretty careless, too, but in line with Walsh's usual modus operandi. He had flown blind before; he had bought, sight unseen, his failed plays in Quebec and the Northwest Territories, banking on hopeful geological reports. Walsh wasn't looking for much, just a good story, something to tickle the imagination of investors back home. And Busang already offered two required elements of a successful gold promotion: an exotic landscape and a pair of determined geologists.

The sell began as soon as Walsh returned to Calgary. Helping out was Barry Tannock, Walsh's old pal from the Ayrex days. "David had called and suggested we get together," says Tannock. "I wasn't doing bugger all. His concept was to be in big-time contact with all the financial constituencies, a lot of brokers and analysts, and retail buy-ers. He was looking for help promoting stock and setting up [an automated telephone] system called Telemagic. There was just Dave, his wife and me. Everyone sort of did everything. I handled the computers, the fax broadcasts, and helped with the press releases." As head of investment relations, Tannock's compensation was $20 an hour and a performance bonus based on the number of shares Bre-X sold.

On 6 May, 1993, the first missive was broadcast from Walsh's basement bunker. "Bre-X Minerals Ltd. is pleased to report that it has entered the Pacific 'Rim of Fire' as a potential gold and base metal miner," the release trumpeted. "We have targeted Indonesia which is particularly attractive by virtue of its geological setting, a favourable investment climate, and political stability." Walsh added that the Busang property "has been tested with 19 holes drilled . . . yielding numerous intersections of +2gm/tonne."

Walsh was putting some heavy spin on results other geologists had characterized as encouraging, but inconclusive. In fact, his story would only get better. He pushed the envelope further by includ-ing de Guzman's quick resource estimate, dropping the two-to-one odds altogether. Without any statistical back-up, Walsh predicted that Busang's "surface mining potential" was one million ounces of gold. He even had a breakdown on operating and capital costs. "Based on these projections, and gold at US$350/oz, the Company's net annual [income] after tax and cash flow would be US$10 million."

More information dribbled out a few weeks later, although the "news" was another rehash of the de Guzman report. "A reinterpre-tation of the existing database which included 19 shallow diamond drill holes" and "a field reconnaissance have been completed by company personnel," Walsh wrote. One of the random chip sam-ples, he noted, indicated a startling 114 grams per tonne of soil. Walsh also repeated de Guzman's sighting of "visible gold." He was taking a risk by repeating the claim; perhaps he realized he'd made a mistake, because the words "visible gold" never again appear in a

Bre-X press release. In the future, investors were told that the gold grains were too fine to be seen with the naked eye.

Tannock was faxing away to the newspapers and analysts, but the PR campaign failed to attract much interest at first. It isn't surprising. Newsrooms and research departments receive dozens of press releases every day; most go straight into the wastebasket. Reporters who depend on promotional "bumf" for story ideas don't last long in the news business. In the event that a press release is spun into an article, it doesn't always result in positive coverage.

In June 1993, a bemused business columnist from the *Calgary Herald* wrote an account of David Walsh and his disastrous record as a mining promoter. Although it briefly alluded to his new Indonesian play, the column referred to a "confession" that Bre-X had once been forced to make to market regulators. The company's stock was "highly speculative." Its properties did not contain "any known bodies of commercial ore." Walsh came off sounding like a hapless clown. "I'm nuts," he said. "I'm looking for the bucket of gold at the end of the rainbow." Stumbling from one loser of an idea to the next, he "snapped back at doubters who badgered him with telephone calls, asking if the company was dead." The writer's conclusion: Bre-X "was no stock for widows and orphans."

It wasn't exactly the coverage Walsh had in mind, especially when he was trying to raise money. He had until the end of the month to come up with McLucas's $80,000 option payment. Then there were Felderhof and de Guzman to consider. Walsh need a few hundred thousand dollars to build some kind of administrative presence in Jakarta and get his two geologists started on an elementary exploration effort. The whole future of the promotion depended on churning out fresh numbers.

Walsh would have to come up with the initial seed money himself. He gathered up half a million of his bargain-basement Bre-X options and sold them at a higher price. "What I did was, my wife and myself had Bre-X stock options that had various strike prices of ten cents, twenty cents, or whatever, and we...got approval from the [Alberta] exchange to up the strike price to forty cents.

And I convinced some close friends and business associates that if they bought the [options] off me, the money would go into an exploration program in Indonesia. And that gave us $200,000 to put in the treasury, which equated to US$150,000. So US$80,000 went for the option on Busang, and the balance was [for] exploration."

Actually, Walsh depended on "certain employees and directors" to buy in. They included John Felderhof, who got an unspecified number of shares. Walsh also sold one and a half million Bre-X shares out of Bresea's account. These moved for about 40 cents each, and had cost nothing to acquire. Pure profit. That summer, Walsh collected over $1 million by free-trading his own stock. At Felderhof's urging, he agreed to pick up a couple more prospects, one on the island of Sulawesi, the other on Sumatra, a large island northwest of Java.

Tannock, meanwhile, was busy working the telephone. "We were promoting very heavily," he says. "All I had to know was what I was prepped to say publicly. David was big on phone contacts, reinforcing the personal touch. He'd get names through newspapers and magazines, people like stockbrokers, analysts, newsletter writers. He got me to organize them on a software system. Eventually we had thirty-five hundred names. It was an efficient system, much better than your average junior resource company. We had information on each contact, with the basic demographics. We'd rate each investor as a prospect, big or small, good or bad."

Walsh and Felderhof hit the road, selling friends inside the investment community on the merits of Indonesia. "We did information meetings to the brokers," Walsh told me during a long conversation in January 1997. "We started in Vancouver. Like, when you're 40 cents, that's where you start. And I was amazed at the people's lack of geographical knowledge. People really didn't know where Indonesia was. So we had to educate them on where it was, the geological potential, sell them on John's talents, and sell them on my talents. So it was quite a job."

In October 1993, the effort slowly began to bear fruit. Trading in Bre-X stock had been fairly active over the summer, and then "all of a sudden we started to get a lot of action," says Tannock. "Somebody was buying stock like wieners at 70 cents."

What caused the stock to jump? Tannock thinks it had something to do with a boost from Walsh's old drinking buddy, Ted Carter. In a late September edition of his investment newsletter, *Carter's Choice*, he wrote about a nifty little company with some "very promising" gold properties. "As most of you know, I rarely trade on the Alberta Stock Exchange and usually recommend against it," Carter wrote. "However, from time to time, information about some stocks crosses my desk that I find very exciting or compelling. Such was the case yesterday with a company called Bre-X Minerals Ltd. I have taken a stock position in this company and perhaps there are some of you who may be interested as well."

Carter didn't mention that he had known Walsh for years, that his pal had a terrible track record in the mining industry, that he had depended on "people who would support his stock in a time of need." In fact, Carter considered Walsh a "heavy drinker" who "never appeared destined for great success." But when he heard about Felderhof's fame, "what he'd done in Indonesia, I thought, 'Geeze, maybe David's finally latched onto a winner here.'"

According to Tannock, Carter's recommendation elicited "a terrific response. The next day we traded 160,000 shares. The bulk of people were buying off of Ted's newsletter. It was a very powerful support." But there was another significant factor at work.

Paul Kavanagh had never heard of David Walsh or Ted Carter, but he liked the sound of Bre-X right away. An executive with American Barrick Resources in Toronto, Kavanagh was always on the look-out for interesting junior exploration outfits. Kavanagh's title, senior vice-president of exploration, was a bit of a misnomer. Barrick didn't explore much; rather, it acquired. The company's CEO, Peter Munk, saw no point fussing around in the forest. Barrick was barely a decade old and already among the largest, most profitable gold producers in North America. Munk's strategy clicked when the price of gold dropped below the crucial benchmark of US$300 an ounce in 1982. Mining fell out of vogue, and Barrick scooped up some excellent properties in Canada and the United States. Its annual production ballooned from 3,000 ounces to 1.6 million ounces by 1993.

Kavanagh's job was limited to initiating joint ventures with smarmy little exploration companies such as Bre-X. It wasn't a very demanding task. Kavanagh was sixty-five, however, and while he still had plenty of energy, Barrick was retiring him. After forty-two years in the business, he was being put out to pasture. But not before he made one final trip.

"My first contact with Bre-X was when I read a write-up in *The Northern Miner*," says Kavanagh. "The results intrigued me, and I contacted Walsh." Kavanagh's interest in Busang proved to be crucial, both in the short term and down the road.

According to another Barrick executive, Bre-X used Barrick to boost its share price. "That's certainly what happened," says the senior vice-president, who met with me on the condition he not be named. "Kavanagh is contacted by a guy called Barry Tannock. He says Felderhof is going to be in Canada, and invites him to meet with them. And so Felderhof and Walsh have breakfast with Kavanagh at the Royal York, and they tell him that Bre-X wants, uh, would welcome an investment on the order of $200,000 as the first step in some kind of ongoing relationship. And as a kind of carrot for this deal, they offer Barrick the right to take control of any of their other Indonesian projects. That ultimately leads to a note from Felderhof in September saying, 'Look, I've been thinking, you guys ought to come to the property in October.' And so from September to October they chat about trip logistics. And as soon as Barrick commits to going to Indonesia, Bre-X starts to publicize it. In early October, they sent out a fax to the press and everybody and their dog saying that a major company will be visiting their property, and it gets picked up by George Cross, a newsletter writer, and the price of the stock goes from like forty cents to one dollar, just on the basis that we're going out there."

Kavanagh travelled to Busang with Larry Kornze, Barrick's geologist in charge of exploration in the United States. Kavanagh liked what he saw. "I subsequently wrote a report late in October strongly recommending that Barrick become involved in Bre-X itself and in the Busang property directly." In early November, he cleaned out his office at Barrick and went to work for Walsh.

The Bre-X file was handed to Kavanagh's successor, Alex Davidson, who was not so taken by Indonesia. Davidson's focus was fixed primarily on South America, where he had more contacts and more experience. But Bre-X persisted. The Calgary company didn't release its first drilling results until mid-December, but Walsh and Felderhof passed some of the data to Barrick in advance. Davidson found it "moderately interesting" but waited for Bre-X's next move.

Bre-X was, "as usual, running out of money," says Walsh. "It was behind the eight-ball, really." According to Barrick insiders, the Calgary promoter began to "rub against our legs." On November 11, Walsh proposed another deal, this time worth $500,000. Barrick would receive approximately half a million Bre-X shares, along with an option to take control of Busang and the two other Indonesian properties. A couple of days later, Tannock called Davidson and suggested Barrick hurry up with a formal response. He added that Bre-X was also talking to a number of Barrick's competitors, including Teck Corporation and Placer Dome Inc., both based in Vancouver. Davidson continued to stall.

Just before Christmas, Bre-X rolled out the initial drilling results. The first two holes encountered weak mineralization. The fourth was stopped short because of a technical glitch, although the early indications were promising. Hole three was a beauty. An eighty-metre length of cylindrical drill core that Bre-X pulled from the ground contained an average grade of three grams of gold per tonne. The concentration peaked at 6.58 grams per tonne. Bre-X shares jumped to $1.20.

Barrick took note. "Standing alone, it wasn't a mine or anything else, but it was certainly a very interesting hole," says the senior vice-president. The company wanted to see more, but Walsh insisted that he was broke, that Felderhof and de Guzman couldn't continue their drilling program without a cash injection. It was time to press ahead and make a deal before Bre-X's stock climbed any higher. By the end of the year, it was trading at $1.75.

On 19 January, 1994, Barrick sent Walsh its first written proposal. Barrick would purchase ten per cent of Bre-X's nine million outstanding shares, for $1.15 million. Barrick would also have an option to buy a majority interest in the Busang property later on. It was a good deal for Bre-X, and a day later, Walsh faxed back a friendly response. "It looked like the deal was pretty much done," says the Barrick source. Walsh wasted no time in sending out a press release to every major media outlet in Canada. Barrick's name appeared in big bold letters at the top of the two-page document. Walsh trumpeted that he was "in receipt of a written proposal" from Barrick, and that further details would be announced pending regulatory approval and the "execution of final documentation." Davidson was furious. He'd warned Walsh not to put out any press release with Barrick's name on it until the two sides had a signed contract. "We don't want someone using our name to flog their stock," says a Barrick executive. "Walsh went ahead and did it anyway."

Adding insult to injury, uppity Bre-X turned the deal down. Felderhof called Barrick's proposition "a bullshit proposal." According to Walsh, Barrick altered its demands and began pushing for a bigger slice of the action. "Its terms changed drastically enough that there would have been nothing left for Bre-X," he said. "It was something like sixty or seventy per cent of our eighty per cent interest in Busang, which was just, you know, pure nonsense and totally unacceptable. I told Peter Munk that I found his team very arrogant. And he said it was unfortunate because without arrogance things probably could have worked out well. Just because you're small, some people think you're stupid."

Walsh wasn't dumb. He'd done a masterful job exploiting Barrick, using its cachet to create a buzz around Bre-X while painting himself as the "little guy" fighting off the nasty Bay Street predator. He would eventually admit to me that he had never wanted to relinquish control of Busang so early into his promotion. "We reckoned that we'd come up with, you know, a two-million-ounce deposit and be taken over by a major, and we'd take our four or five dollars a share and that was a two-year exercise, okay?" Although he'd trumped Barrick this time, Walsh would regret the day he ever got mixed up with Peter Munk.

Paul Kavanagh opened another door. The diminutive geologist had no plans to fade into the background after leaving Barrick. In March, he formally agreed to join Bre-X as an "outside director" working behind the scenes to find funding for the Busang project. Kavanagh would not draw a salary but he did pick up several hundred thousand free Bre-X options.

Kavanagh wandered next door to Loewen Ondaatje McCutcheon Ltd. (LOM), a small brokerage specializing in underwriting new mining ventures. LOM's chief gold analyst was Robert Van Doorn. He remembers being impressed when Kavanagh came calling, early in 1994. "The fact that Paul was high on this Indonesian play gave us a good feeling," says Van Doorn. "He was a senior geologist with a good reputation. He was a very important factor in getting people to take Bre-X seriously. He could tell people what the geology looked like, what the company planned to do, how much it would cost, that kind of thing."

Van Doorn, a native of Holland, had spent months in Indonesia. His wife grew up in the former Dutch colony, and his in-laws still lived there. "I knew about the Australian exploration boom," he says. "When Kavanagh came around with his maps of Busang, I was interested, sure. He had a good concept. It was the right place, and Bre-X had set sensible targets. We discussed getting involved and the consensus here was that it was a high-risk play, but the rewards could be very high."

Self-interest played a role, as it always does when a brokerage sells shares in a company. Typically, an underwriter receives a five to ten per cent commission on every special offering it manages to sell. Sometimes, it receives additional payment in the form of special warrants such as stock options. Bre-X asked LOM to manage a private placement, a quick and easy method of raising money. Private placements, which are very popular among new mining start-ups, do not require companies to file a prospectus with market regulators. This means a promoter and his brokerage can say virtually anything they like about their company; there is little if any regulatory scrutiny. Shares are typically sold in large blocks to a select group of individual and institutional investors, ones with deep pockets and a penchant for high risk. LOM, with its extensive contacts in Europe, was the perfect firm to handle the offering. The brokerage agreed to underwrite a $4.5-million private placement and had no trouble moving three million shares in two weeks, pocketing approximately $270,000 in the process. Most of the shares were placed in European banks.

Bre-X kept popping holes, well into the spring. By May, Bre-X reported big concentrations of gold, up to almost thirteen grams per tonne, starting right at the surface. De Guzman's crew kept moving south, towards the edge of the Busang concession. When Bre-X extended its claim into an abandoned forty-five thousand hectare concession adjacent to the original site, the numbers started pouring in. "The Central Zone pales in comparison to the potential of the Southeast Zone," Felderhof wrote. "The Busang project in its entirety has the potential of becoming one of the world's great orebodies."

It was a shocking prediction, based on fuzzy geological theories and a few dozen holes. Over the next twelve months, Busang ballooned to giant proportions, from one million ounces, to six million, to forty million ounces of gold. No one asked Bre-X to prove it.

7

JUMPING IN

*National brokerage houses do not finance junior
companies without doing a lot of due diligence. This
stamp of approval is a signal that what Bre-X has been
saying all along has a lot of merit to it.*
  – Brian Fagan, publisher, *Asian World Stock Report*,
    December 1995

*People are starting to believe that what we have been
saying is true.*
  – Paul Kavanagh, 18 August, 1995

B RE-X COULD DO NO WRONG. Walsh, Felderhof
and de Guzman were engineering the hottest
story in business. Walsh and Felderhof hit the
road, spreading the Busang gospel up and down the west coast. They
put two new drill rigs on the site and recruited new geologists to
scout the fringes. De Guzman called on a handful of friends from the
Philippines to help run the operation at Busang. Canadian and
Australian geologists were sent to Bre-X's two other far-flung mining
concessions, on Sulawesi and Sumatra.

Soon the company had six drill rigs at Busang and was pulling
more core in a single week than most exploration outfits would in a
month. Bre-X punched another hole and — bang — added another

five million ounces of gold to the pot. To average observers who knew nothing about mining, it was simply great fortune. They may not have realized that Bre-X was a frothy promotion. No one bothered to look deep into Walsh's past. Felderhof's statements went unquestioned, and the numbers kept growing.

The experts should have known better. The results were too good, and came too fast. Robert Van Doorn stopped believing, saying that "at one stage, I started telling Bre-X it's unlikely that the holes would get better and better. They always sounded a bit wild." But Van Doorn helped heighten the public's awareness with his first research report, distributed in late July 1995. He rated Bre-X a "buy," which isn't surprising, since his firm had just handled the company's first private placement and was angling for more business. Van Doorn predicted its shares would reach $4 within twelve months. "The next few holes," he wrote, "are expected to be very good." And they always were.

In time, Bre-X would have Bay Street in its pocket. Walsh could say anything, and the analysts would jump. The biggest brokerages in Canada soon followed LOM's lead and piled into Bre-X, lauding its management, parroting its propaganda, legitimizing the entire effort with their blessing. The bigger Busang became, the more shares they traded. It's estimated that in three years, Canadian brokerages made approximately $116 million off Bre-X, just in commissions. That doesn't include the millions some firms made flogging juicy private placements. And it doesn't take into account commissions from $2 billion worth of stock traded among other Canadian junior explorers, including the thirty-odd outfits that gleefully followed Bre-X into Indonesia, hunting for elephant gold deposits in the rainforest. New companies with evocative names such as Borneo Gold Corporation, East Indies Mining Corporation and South Pacific Resources Ltd. suddenly appeared on the junior stock exchanges, touting their proximity to Busang.

Raising expectations wasn't just a promoter's job. Time and time again, the analysts, whose ranks included experienced geologists, took the task upon themselves. They're supposed to work at arm's length from the brokers, who act for small retail investors and large institutions. Analysts are the senators of finance, there to add a bit of sober second thought to investment decisions. Their role is to scrutinize, provide background on industries, and rate a company's chances for success, weighing all the odds. But caution took a back seat to personal gain. Unlikely as it should have been, Bre-X became a sure thing.

As Bre-X's stock continued to rise, more pundits and analysts began clamouring for an audience with Walsh, Felderhof and Kavanagh. Many threw judgment out the window. Michael Schaefer, publisher of the *Global Gold Stock Report* newsletter, was a huge Bre-X fan. He wrote a number of positive reports about Busang. Bre-X staffers received his US$169-a-year newsletter and distributed copies to reporters and other analysts. In November 1994, Schaefer reminded his readers that "the foremost and primary rule of investing in the junior resource industry is to have confidence in the people behind the company." Somehow, he concluded that Walsh, "the brains behind Bre-X," was "one of the sharpest executive officers that we have ever encountered in this business. He is also one of the most honest and dependable."

Reached at his office in Buffalo, Wyoming, Schaefer laughs and says he "can't remember" what led him to such a glowing assessment of Walsh's management abilities. In fact, he never met the man, nor anyone else from Bre-X. Schaefer dealt with Walsh over the telephone. They spoke often, he says, enough to give him the impression that "Walsh was the weak link" and that "he had a drinking problem." Oddly, he never expressed any doubt about Walsh in his newsletter. (As a matter of "principle," Schaefer buys every stock he recommends. He rode Bre-X all the way from $1.20 to $250 before cashing in.)

Dorothy Atkinson wanted to make money. A twenty-year veteran of the mining industry and with a doctorate in economic geology, Atkinson used to advise companies how to develop mineral deposits in western Canada and the United States. Early in 1994, she switched to the investment business, joining Pacific International Securities in Vancouver. "I was a penniless person working for salary all my life," she says. "I wanted to find something to put me on the map."

Atkinson was "looking around for a story I could be successful with, starting with a geological area that had been overlooked. Everyone was onto Africa and South America, but not Indonesia. Bre-X just stood out." In August 1994, she started covering the Calgary company. Her first report, which delivered a "buy" recommendation, was essentially a rewrite of Bre-X press releases, lacking any independent analysis whatsoever. "I had seen de Guzman's reports," she says. "One kind of assumes that it was all right. You just go along with the story."

Pretty soon, Walsh was all over her. "He was exceptionally good to me," she says. "He always came to see me when he was in Vancouver. He was a broker. He understood how to play the game." She continued to write positive reports, even after moving to another, larger firm, Whalen Beliveau and Associates Inc. Eventually, Atkinson was rewarded with a $29.4-million private placement in Bresea Resources, which owned twenty-five per cent of Bre-X.

"We had approached Walsh and asked him if he needed any financing," says Atkinson. "Right then and there, he gave us a mandate to go find the money. The deal was made through me. As an analyst, you have to make relationships and bring business to the company you work for. It's up to you to be profitable. So I was ecstatic. Unfortunately it was all based on lies."

Only a few sceptics stood apart from the sycophants. Most of them were on the buy side. Paul Stephens, a San Francisco-based mutual fund manager, was approached by Bre-X in early 1994. Unlike the vast majority of his colleagues in the investment industry, Stephens ran some quick background checks on Walsh and Felderhof, which left him completely unimpressed. Walsh persisted, however, and showed up in Stephens's office with his wife, Jeannette. "He had this massive pot belly," Stephens told *Fortune* magazine. "Our checks on him were just awful. He had a history of filing for bankruptcy. And then, when you met him in person, he was a slob. Bottom line: All you had to do was meet him and you wouldn't have bought the stock."

Nevertheless, Stephens's geological consultant continued to monitor Bre-X. "It was *the* hot topic. I constantly had to defend the fact that we didn't own it. I was wonderfully intrigued," Borden Putman told *Barron's,* "but I couldn't get the story filled out to my satisfaction. I participated in two or three conference calls they had, though, and I asked a lot of pointed questions, which they never answered." Only a bad analyst or money manager would recommend investing in a junior mining company by relying on some "pat story" floated out by its management. "They tell you a lot of things, and then I go out and try to verify."

Other mutual fund managers later tried to make hay out of the fact that they avoided Bre-X, even if it meant stretching the truth. James Turk, a strategic advisor to the New York-based Midas Fund, appeared on American television, explaining that his company had considered buying Bre-X about half a dozen times, starting when the company was trading for less than one dollar. But "a number of red flags kept coming up," which persuaded Turk and his partner not to get involved. "It shows our selection process makes sense," he told one reporter. When I contacted him, Turk refused to say what the warning signs were, "due to sensitivity in the marketplace." He did allow that he had conducted "extensive due diligence" on Bre-X and its management, didn't like what he saw and passed.

But documents filed with the Commission des Valeurs Mobilières du Quebéc show that Turk's Midas Fund was a major participant in the Bresea private placement handled by Dorothy Atkinson and her firm, Whalen Beliveau. Walsh had promoted Bresea as a cheap way to get involved in Bre-X. Bresea was easily its largest stockholder, with approximately five million shares. One month prior to the private placement, Atkinson noted that "Bresea is more than a holding company. It shares management with Bre-X and has participated in the financing that led to the discovery of Busang. The company is a

joint venture partner of choice for Bre-X and will continue to participate in exploration."

Confronted with evidence that the Midas Fund bought 200,000
Bresea shares, one-tenth of the entire private placement, Turk
became defensive. "Bresea was not Bre-X," he sputtered. "We only
owned it for a short time. I didn't like the look of the chart so we sold
it. By the end of 1996 it was gone." Finally, Turk admitted he really
didn't know much about mining. "My partner has more experience
in geology than me," he said. "Neither of us has a geological background."

Barbara Horn doesn't understand geology either. The Calgary-based
broker with Nesbitt Burns initially depended on Bre-X to educate
her about Busang. Although she'd been burned by Bre-X in the past
and had been "mad" at Walsh because he "hadn't done his homework," she liked the new promotion. Felderhof particularly impressed
her; he seemed passionate and had lots of experience in Indonesia.
"He knew his geology," she says. "And David really understood the
brokerage business. It was almost a perfect combination." As the
months passed and Bre-X began to roll out the big numbers, Horn's
scepticism began to fade. "I had lost faith in David when things
didn't work out up north," she says. "But if you can believe in a person, what they say, then you can always turn around and change
your position. And I believed him."

Horn bought back in at 70 cents, and watched her $7,000 invest-
ment quadruple in a year and a half. Suddenly, she had nothing but
fondness for Walsh. "He doesn't forget his friends," she said in
December 1996. "He's genuine in that he doesn't play with you at all.
He says what he thinks, in a simple way. He doesn't make things
complicated. He always says the positive thing." Horn also liked
Jeannette, who "worked her buns off." One Christmas, she gave the
Walshes a silver ball with gold sparkles to hang on their tree. "I can't
say enough good things about them," she told Sean Silcoff, a reporter
with *Canadian Business* magazine. Her opinion would soon change.

In April 1995, Bre-X announced the first drill results from the new Southeast Zone, which the company had acquired a year earlier. Tests conducted at a local assay lab indicated that the first hole showed three grams of gold per tonne, over a depth of almost two hundred metres. This was a very good score, and Bre-X shares zoomed from $2.50 to $4. In July, the company reported that three more holes, spaced well apart, had hit similar quantities, and the price doubled again. Bre-X set a new resource target of six to eight million ounces.

This went well beyond any previous projections. But Felderhof and de Guzman had a theory. The gold-bearing diatreme dome they thought they had found in the original Central Zone was actually the tail-end of a much bigger system to the south. The two geologists distributed a technical report, explaining how it all worked. To the layman, however, "dome lithologies" and "massive crystalline quartz-hornblende dacite" and "basalt-andesite-dacite-rhyolite plugs and dikes" meant nothing.

Horn turned to her firm for help. Nesbitt Burns, Canada's largest brokerage, has a large research team at its disposal. Its chief gold analyst, Egizio Bianchini, was considered one of the nation's pre-eminent precious metals specialists, according to an annual peer review. Horn came to rely on Bianchini for information about Bre-X. He had a better relationship with Walsh and Felderhof than any other analyst on the Street. Too close, perhaps.

Bianchini was a number cruncher, not a geologist. He had majored in geology at the University of Toronto, but never worked in the field. Instead, he obtained an MBA from the University of British Columbia in the early 1980s and went to work for Nesbitt, where he quickly rose through the ranks, carving a niche for himself in mineral economics. It's a dry, academic field, one that Bianchini believed was terribly underemphasized in the mining industry. He loved to proselytize about the significance of gold in the global market. Unlike many of his colleagues, enraptured by a prolonged bull run in the securities market, he rejected the view that the metal was passé. In speech after speech, he defended its "traditional role," even as people around the world were "openly challenging its monetary value" and sidling over to alternative, higher yielding forms of investment. Gold, he liked to say, "is not just a commodity. In the history of mankind, it has always represented a store of real value."

Bianchini may have seemed like a conservative, but he lost restraint when it came to small exploration companies. He had an opportunity to make a name for himself, as several of his colleagues at Nesbitt Burns had done when the Voisey's Bay nickel discovery was made and they jumped on the bandwagon, early. Bianchini was thrilled when more high-risk junior outfits began scouring the globe for new gold deposits. Canadians should abandon their "reluctance to invest abroad in mining ventures." It only made sense, he said. "Record low interest rates caused a tremendous flood for liquidity into the capital markets," he noted in 1994. People "needed investment vehicles, and the mining industry provided them.... Exploration successes began to be recognized in Chile, Brazil, Venezuela, and other undesirable places."

Inevitably, Bre-X caught his eye. The company was sparking another regional exploration boom, this time in Southeast Asia. Nesbitt Burns had participated in a private placement for Bre-X in the spring of 1994, and there was plenty of buzz about the company inside the brokerage's Bay Street headquarters. In September 1995, Bianchini hopped on the bandwagon.

Bill Stanley was surprised to see Bianchini at the big Royal York meeting that month. Officially, the Nesbitt Burns analyst had only been covering Bre-X for a couple of weeks, and yet there he was, centre stage with Walsh and Felderhof, boiling with excitement, almost manic. "Egizio was running around, grabbing clients by the hand, taking them into corners of the room and jabbering away. You have to appreciate what it meant to hear Egizio talking up a company like that," says Stanley. "He was one of the better analysts around. He had tremendous clout. His recommendation could add five to ten per cent to a stock. He was in love with Bre-X."

Bianchini's first report, dated 5 September, 1995, urged investors to buy stock. "After reviewing all the drill data and all available cross-sections for Busang," Bianchini wrote, "we have developed the view that the project is likely to evolve into a world-class mining operation." The geology at Busang "displays many, if not all, of the attributes required for a potentially economic gold reserve. . . . Management is very competent technically." He set a target share price of $21.

Barely a month later, he revised his forecast to $50 and gave Bre-X a premium "S-5" buy rating. Juicy new results on three holes in the Southeast Zone had just come in; Bianchini now advised that "Busang will become one of the elite world-class orebodies, both in size and profitability." The numbers being thrown around, he gushed, were "eye-popping." If Bre-X kept churning out the results, Busang would "likely contain the world's single largest deposit of gold."

For the next year and a half, Nesbitt led the pro–Bre-X brigade on Bay Street. In the absence of independent confirmation, any exploration play must be regarded as speculative. Yet Bianchini abandoned all doubt, referring to information gleaned from Bre-X as "fact." He called his resource estimates "conservative" although they tended to be higher than even his most bullish colleagues. Some allege he was reckless. Certainly Bianchini threw caution to the wind, telling one wide-eyed *Globe and Mail* columnist that "good properties tend to be found by guys with checkered pasts, guys who can hustle up money. Don't be scared of promoters with lousy track records."

Lawyers now working for a group of disgruntled Bre-X share-holders claim Bianchini ought to have known that the company was making misleading statements. They suggest he received special favours from Bre-X, such as exclusive access to management and to Busang, helping him stay ahead of the competition. In any event, Bianchini displayed an uncanny ability to forecast huge resource estimates for Busang, well ahead of anyone else.

A spokesman for Nesbitt Burns says the firm's analysts deal only with "publicly available information. [Acting on] insider informa-tion is illegal." But in early November 1995, two months after he initiated his Bre-X coverage, Bianchini was afforded a rare trip to Busang. He was, in fact, the first analyst ever allowed on the site, and he came home impressed. During a breakfast meeting with one Bre-X investor, Bianchini reportedly confided that he had some unpublished drilling results from the Southeast Zone. He couldn't share them, due to "securities regulations," although he felt comfortable estimating that Busang contained thirty-four million ounces of gold. This was almost double what Bre-X was predicting, but Bianchini did not explain the discrepancy. He wrote another bullish report on November 17, with his thirty-four-million-ounce forecast. Three days later, Bre-X shares hit Bianchini's $50 target. He moved the figure up another notch, to $70.

According to Cam DeLong, a senior geologist with South Pacific Resources Corp., an exploration company based in Vancouver, mining officials do share confidential information with analysts, in order to "get rumours going on the street. But you really shouldn't be doing that," he says. Besides, he adds, "a company that has real results will generally only disclose them publicly. Why crank up the rumour mill if you've got the goods?"

Bianchini continued to rewrite his forecasts. Another private placement was in the works; this time, Bre-X was looking for big money. The company fed its numbers to Bianchini, and he boosted his estimates. In January 1996, he repeated his "conservative" prediction that Busang contained more than thirty million ounces of gold, while increasing his target share price to $110. Nesbitt Burns led in a $30-million private placement, selling 250,000 Bre-X common shares at $120. Meanwhile, the brokerage was buying and selling more Bre-X shares on the retail market than any other firm on Bay Street.

The more Bianchini boosted Busang, the more profits rolled in to Nesbitt Burns. Why he never questioned its credibility, even in the last days before the meltdown, is still being debated. Some have suggested Bianchini was in so deep, he couldn't step out. Later, Bianchini allegedly confessed to a group of colleagues that he had always assumed PT Kilborn Pakar Rekayasa, a Canadian-owned engineering company based in Jakarta, had routinely analysed Bre-X's drill core. Walsh said as much when he declared that Kilborn, a subsidiary of Montreal-based engineering giant SNC-Lavalin Inc., had provided "metallurgical testwork" on Busang drill samples.

But Kilborn denied it, insisting the "scope of its mandate from Bre-X relates to resource studies and modelling." The company said it merely took numerical data presented by the Indo Assay lab in Balikpapan, fed it into a computer and came up with a "resource estimate." Kilborn was, in fact, hired by Bre-X to prepare a major feasibility study of the Busang project. Its job included designing a suitable mine, detailing employment and equipment needs, accessing social and environmental impacts, and determining capital and operating costs. The engineering firm's intermediate report, the size of two Toronto telephone books, also contains detailed metallurgical studies of the Busang drill core.

These studies mentioned some troubling "discrepancies" regarding test results from the Central Zone and recommended that "a larger pilot type trial be undertaken" to "gain further confidence in the resource assays." A metallurgical company contracted by Kilborn suggested it "drill a parallel hole(s) directly adjacent to an existing hole" and have the entire core analysed by a third party. "The amount of recovered gold would then be compared with the amount of gold predicted by the parallel hole." The contractor later recommended the same procedure for the Southeast Zone. Yet Bre-X ignored the recommendation, explaining that "all the drilling had to focus on expanding the reserves."

Conveniently, the Kilborn report was kept under wraps. Meanwhile, the firm's resource estimates just kept getting bigger and bigger, beyond anyone's wildest expectations. From August 1995 onward, all of Bre-X's new resource estimates bore the Lavalin/ Kilborn seal of approval. Bre-X began running an official-sounding tag line after every estimate, noting that Kilborn's work was based on the "Principles of a Resource/Reserve Classification for Minerals by the U.S. Bureau of Mines and the U.S. Geological Society." No one had a clue what this meant, but it lent legitimacy to the whole Busang effort.

Bianchini was convinced. If Kilborn said the gold was there, then Busang must be legitimate. He wasn't the only one fooled. "Kilborn's name was absolutely key," says an official with Indomin Resources Ltd., a Canadian exploration company with property next to Busang. "My guys in Indonesia figured the whole thing was a scam, that Felderhof was up to something. But when we saw Kilborn's name on those resource estimates, we all changed our minds."

No one dug up the Kilborn/Lavalin feasibility study — with its damning metallurgical reports — until it was too late. Bianchini can't explain why he didn't ask to see them. He was muzzled by Nesbitt Burns once the lawsuits began pouring in. "I'd like to talk about it with you, but I can't," he told me, during a brief telephone conversation. "All I will say is, don't believe everything you read." It's a remarkable comment. Bianchini's reports convinced hundreds, perhaps thousands, of investors to buy into Bre-X. Sadly, the irony seems lost on him.

8

CRACKS

I have a personal mission to try and bring corruption out of the closet. But nobody seems to want to talk about it.
— Bob Parsons, former head of Price Waterhouse's world mining group

THE CRACKS WERE SHOWING well before Bre-X was ever exposed, but Bay Street was blinded by greed. As the Busang promotion roared into overdrive, dozens of other explorers sniffed the money and went after it. Some of the juniors were serious about finding gold. Led by weathered veterans, they were part of a professional Canadian brigade that had been operating overseas for the better part of a decade. A few even had the goods. Arequipa Resources, a junior exploring down in Peru, discovered a multimillion-ounce gold deposit and struck a huge deal with the acquisitive Peter Munk, selling Barrick two properties for $1.1 billion.

That was a score. But most of the juniors didn't have a hope of finding anything. Some were pure froth, passed off on investors by salesmen and financial advisors who, in turn, were under pressure to close deals and bring in the money. Canada's big brokerages were underwriting lucrative private placements and new share issues at a record-setting pace. Exploration companies were the hottest investment plays in 1995 and 1996, thanks in large part to a relative shortage of opportunities in other sectors. Desperately searching for places to park almost $650 billion worth of investment capital, Canadian mutual fund and pension fund managers were ready to look at anyone with mining properties in Africa, Borneo and other exotic locales.

"We were in a situation where a lot of people were buying into a small industry over a short period of time, and investors had trouble separating the winners from the losers," says gold analyst Vic Flores. "We were bound for a correction." Bre-X, he says, was "the grand finale to all the pain and suffering in the sector."

When Cam DeLong joined South Pacific Resources, the company had a few concessions scattered around Kalimantan. It hadn't punched a single hole in the ground, but that didn't matter. South Pacific raised $14 million in a private placement — based on nothing. "I was amazed," says DeLong. "I didn't think you could get money so easily." Harold Jones, chief geologist with Pacific Amber Resources Ltd., another Vancouver-based junior, remembers raising $15 million in a private placement, "all on the hype of Busang. We just tagged along with it." Pacific Amber's stock, which had languished well below a dollar for years, suddenly shot up to $10. "It was absurd," says Jones. "You could not justify it. The whole thing was nuts."

Rather than step back and call it for what it was — a mindless speculative frenzy — some of Canada's top investment firms took full advantage. They traded up on bad information. They bought personal shares in penny-stock companies and then purchased large blocks — at higher prices — on behalf of their clients. They became de facto insiders, taking management and ownership positions with exploration outfits and then flogging their stock on the market. They didn't check the facts. At best, they were involved in some serious conflicts of interest, even wilful neglect. But Canadian securities regulators, hobbled by inadequate funding, could barely enforce their own laws. Penalties for the most severe transgressions, such as insider trading and stock manipulation, were laughably weak, mere

slaps on the wrist. Voluntary codes of conduct were routinely
ignored; the pros on the street argued they hampered their competi-
tive edge. Investment houses looked the other way as analysts, brokers and mutual fund managers became active players in the mining
bubble, cashing in on the boom. There was nothing to stop them
from leading their clients into minefields.

The bombs began exploding in early 1996. Timbuktu Gold Corp.
was the creation of Yorkton Securities Inc., a large Toronto-based
brokerage with offices worldwide. Timbuktu ran up a $650-million
market value by pretending it had found gold in distant Mali, a
riotous country in Western Africa. The brokerage established control of a broken-down penny stock called Choice Software Systems
Ltd., languishing on the Alberta Stock Exchange, and in late 1995
paid off its debts, changed its name and geared up for a new promotional shtick. In December 1995, Timbuktu bought a mining concession in Mali from a shaky Las Vegas promoter named Oliver
Reese.

His address should have been the tip-off. Reese was bad news.
He'd already been nailed for making false claims about another
alleged gold deposit in Mali, but this was a little bump in the highway the Yorkton boys never disclosed to securities officials. Reese
had a new wonder, a piece of desert he'd bought a few months earlier
for US$225,000. He sold it to the Yorkton team for US$1 million and
ten million Timbuktu shares. He also became Timbuktu president.
Yorkton and Cannacord Capital Corp., a penny-stock brokerage
operating out of Vancouver, were the second and third largest shareholders. Just after New Year's, Timothy Hoare, the British brokerage, sold a four-million-share private placement to "unidentified
Europeans," raising $4.8 million. None of it made any sense, but
Timbuktu now had money.

The company moved a "team" onto the Mali property. A couple
of holes were reportedly drilled. In mid-April, Reese received word
from the lab — the drill core was fabulous. It contained huge grades

of gold, as high as 187 grams per tonne. According to an official at Timothy Hoare, "someone" wrote a scintillating review that "somehow" was posted on the Internet. "Timbuktu Gold stands on the threshold of a major gold discovery. [It] looks to have hit the elephant on the first shot."

The market rushed at it. Timbuktu shares soared 180 per cent, even as prudent mining analysts warned the company's statements were nonsense. The results were inconceivable. Yorkton handled fourteen per cent of all trading conducted during the run-up, more than all other brokerages. Cannacord was next with six per cent.

"It's the Bre-X phenomenon," gushed one of Timbuktu's directors. He got that right. Reporters in Calgary rushed to get David Walsh's comment. Timbuktu, he said, was another example that showed why the Alberta Stock Exchange was a great place to run a mining promotion. The requirements were "stiff," he said, but any "guy with an idea can go to the Alberta exchange, and if it's legitimate, they'll listen."

Timbuktu was a scam. A simple background check on Oliver Reese would have scared anyone off the company. Ed Waitzer, former head of the Ontario Securities Commission, said Yorkton and the ASE had an obligation to look into Reese's past. But ASE officials didn't learn about Reese until a business reporter sent them some press clippings. Horrified, the exchange ordered Timbuktu to stop trading and sent an independent group of scientists to Mali to double-check the results. The conclusion: Sample bags of crushed Timbuktu core had been salted with alluvial gold. Timbuktu was a bust. When trading recommenced months later, the share price crashed through the floor.

Meanwhile, at Yorkton Securities, business continued as usual. The brokerage had — wittingly or not — backed a sour promotion. At the very least, it had shown terrible judgment by hooking up with Reese. As far as company officials were concerned, however, the brokerage had done nothing wrong. An internal review of Yorkton's trading records revealed no irregularities, said the firm's chairman. In his mind, the case was closed.

The trash kept coming. Cartaway Resources Corp. leased plastic garbage containers before eight mutual fund managers from First Marathon Securities Ltd. transformed it into a mining promotion. Listed on the Alberta Stock Exchange, Cartaway was set up to exploit the exploration boom at Voisey's Bay, Labrador, where Diamond Fields had found its multibillion-dollar nickel deposit. First Marathon's "Group of Eight" raised a quick $2 million and began acquiring a group of claims in the same area. Controversial promoter John Ivany, banned from trading on the Vancouver Stock Exchange in 1991 thanks to disclosure violations, was recruited as president. Michael Stuart, a First Marathon executive, moonlighted as Cartaway's chief financial officer. Together, the two men managed to attract a number of impressive institutional investors, including the largest mutual fund company in the world, Boston-based Fidelity Investments, which took a ten per cent stake in Cartaway.

In late April 1996, just as Timbuktu was unravelling, Cartaway reported it had found high copper grades at one of its Voisey's Bay properties. The news was based only on a quick eyeball estimate but the stock took off, from $2 to $26, much to the delight of the First Marathon men, who had millions of cheap shares in their own trading accounts. Stuart sold 145,000 shares within a week, reaping $1.4 million. Cartaway's exploration boss sold 50,000 shares, grossing $414,000. Then the assay results came in, reading a fraction of the copper Cartaway had suggested in its reports. It was "an honest mistake," Ivany explained. But the damage was done. Cartaway's stock dropped back to $2.

Predictably, First Marathon found "no evidence of market manipulation or insider trading" by any of its brokers. The company did, however, introduce new guidelines prohibiting employees from running penny stock companies and acting as stock promoters. The move seemed logical, albeit long overdue, yet it offended a number of industry players. "The real blame [for the Cartaway crash] must fall squarely at the feet of investors and speculators who chased Cartaway's stock mercilessly," complained the Vancouver brokerage of Lee, Zaunscherb, and Associates. "The whining should end."

But what happened to the notion that brokers must act in the best interests of their clients? In theory, brokerages and mutual fund companies monitor their employees' personal trades to prevent conflicts of interest. But Canada's investment industry is self-regulated, so there's no way to ensure compliance. Codes of conduct drawn up by umbrella associations such as the Investment Funds Institute of Canada are strictly voluntary. Harold Hands, chairman of IFIC's ethics committee, says there's nothing to prevent mutual fund managers and employees from ignoring certain sections, including those that deal with personal investing. As a result, observers have concluded that promises to clean up the industry amount to little more than lip service.

This view was enforced when another mining promotion went bad. Once again, brokers were seen fleeing the scene. Consolidated Brenzac Development Corp. was a rudderless VSE-listed casualty until it was acquired by a group of Vancouver promoters. They tried to turn the company into another Voisey's Bay vehicle, but that didn't sell, so management decided to jump on the Indonesia bandwagon. They raised $7 million in a private placement; among the customers were Robert Disbrow and Rob Hartvikson, two brokers from First Marathon's Vancouver office. A former Toronto stockbroker named Wayne Wile was also a buyer. Wile was once convicted of offering illegal sales commissions to friends in the industry, but the ruling was dismissed on appeal due to improper use of wiretap evidence.

In April, Consolidated bought a handful of "unusually good" mining concessions in Kalimantan and, a week later, changed its name to Borneo Gold. Six months later, Borneo claimed it hit had pay dirt. "An extensive area of bedrock containing abundant coarse gold has been discovered in West Kalimantan," the company crowed. During a two-day walkabout, a group of geologists had found gold lying all over the place, in every form imaginable, including "porous blebs, nuggets, plating and dust." According to Borneo, "Bonanza grades can be expected."

It was ludicrous; Borneo didn't have a shred of evidence to back up its claims, just a visual inspection of a few grab samples. To some, it smelled a lot like Cartaway. One veteran stockwatcher suggested that a powerful promotion was at work. But Borneo's announcement had the desired effect. Shares shot from $2.55 to $10.05 before the VSE's embattled director of market surveillance halted trading and demanded to see some assay results. Borneo director James Anthony took offence. "It's hard to mistake gold," he said. "We have six geologists. [They] viewed the samples on site and are telling us that this is abundant gold, so we felt that we were safe."

One month passed. No news. Then Borneo issued another statement, a little more conservative than the last, but still claiming the property was riddled with visible, free grains of gold. Samples assayed at a lab in Toronto revealed high grades. The VSE allowed trading to resume but Borneo shares immediately hit the skids. It's not clear when Disbrow and Hartvikson bailed out, but a colleague of theirs, Francis Manns, continued to back the company. A metals analyst with the securities firm Marleau Lemire Inc., he remained "optimistic Borneo Gold has made a major gold discovery" and predicted an eight-million-ounce deposit, minimum. The market didn't share his confidence, however; Borneo continued to plummet. A year later it was stuck in the two-bucks-a-share zone. Most of the news that trickled out of headquarters was related to bushels of new stock options for directors and key employees.

Why was Manns so bullish on the company? It's hard to say. He can't be found at Marleau Lemire any more. The firm's brokerage division was shut down in early 1998, after piling up huge trading losses in the junior mining sector. Manns has a new address, in First Canadian Place. He's the president and CEO of Borneo Gold. And he doesn't return telephone calls.

Canadians are not generally known as risk-takers, but the latest rash of one-month mining wonders demonstrates once again their curious and long-standing susceptibility for bad promotions. Marc Cohen, a respected gold economist with Paine Weber in New York, says American "sell-side" analysts seldom follow junior explorers. "Down here, the outlook is more long-term. Personally, I deal only with companies which are actually in the business of producing gold. I don't talk to people like David Walsh every day of the week, unlike my Canadian colleagues."

At the very least, says Cohen, Canadian analysts can do a better job of scrutinizing junior exploration outfits before recommending them for investment. They need to determine the risks based on a careful assessment of the people involved, not just on the potential of a property. Investors can protect themselves from short-lived promotions by doing their own homework, rather than simply relying on reports from their brokerages. Or they can simply follow Cohen's advice and restrict their investments to mining companies already in production.

But that raises other dilemmas. Canadian miners have become increasingly outward-looking in the 1990s, spending record sums exploring and developing mineral properties abroad. Cash-strapped countries in South America, the former Soviet bloc and parts of Asia have opened their doors, signalling a new era of mining activity; it's estimated that seventy per cent of the earth's surface is now available for mineral exploration. But while the new activity helps boost local economies, providing jobs, housing and infrastructure, it has also led to grotesque practices that would never be countenanced within Canada. Eschewing ethics for profit, some Canadian companies working overseas have ignored local concerns, wreaked environmental chaos and made underhanded deals with corrupt military regimes, even funding armed rebel movements in Africa.

Canada's worst mining imperialists are bent on exploiting the resources of weaker, unstable nations, while giving as little back as possible. Heading the new wave is Robert Friedland, co-founder of Diamond Fields and one of the most notorious mining promoters Canada has ever produced. A master of intimidation, Friedland is known to bully and threaten people who stand in his way to a business deal. According to David Baines, a Vancouver-based mining reporter, "He is an expert at the craft of launching junior mining companies with deals that effectively transfer the otherwise long odds of mineral exploration to public investors."

Friedland's reputation was severely tarnished before Diamond Fields discovered the Voisey's Bay nickel deposit and made a lot of analysts, brokers and pundits rich. Half a decade earlier, a Friedland-owned company was involved with one of the worst environmental mishaps in American history. The Summitville gold mine in Colorado was transformed into a toxic pit after cyanide leaked from poorly designed containment devices. Cyanide was also illegally discharged into a tributary of the Rio Grande River. The Environmental Protection Agency laid thirty-five charges against Friedland's company, alleging that it had committed infractions "based on greed," and that a mine worker had falsified reports about the poisonous discharges. Friedland denied any responsibility, claiming he had no personal knowledge that the mine was faulty when he purchased it. Nevertheless, he put his company into bankruptcy, leaving American taxpayers to pay for the US$100 million clean-up.

In 1993, Friedland hooked up with Montreal-based Cambior Inc. to develop a gold deposit in Guyana, South America. Friedland's Golden Star Resources Ltd. is now run by former Canadian football star Dave (Dr. Death) Fennell and lists former federal cabinet minister Don Mazankowski as a director. It has a thirty per cent interest in the Omai mine in Guyana. Cambior, which boasts former Canadian trade ambassador Gordon Ritchie as a board member, owns sixty-five per cent of the mine. The largest open-pit gold operation in South America, the Omai mine employs a thousand local people and, in the words of one environmental activist, is run like a "military industrial camp." Protected by armed security, it is managed by Norman McLean, former chief of staff with the Guyanese army.

In April 1995, Cambior asked the Guyanese government for permission to discharge cyanide waste into a local river used by local people for drinking, cooking and washing. An effluent pond next to the mine was in danger of overflowing, according to Cambior. If the government refused, the company would cease operations.

It was a serious threat. The Omai mine was the largest commercial enterprise in the country, responsible for almost half of Guyana's entire gold production. Still, the government could not go along with the plan, since tests demonstrated that the effluent's toxicity level was seventeen times higher than the U.S. safety standard. Cambior would have to find another way to dispose of the tailings. One month later, a spill dumped cyanide-laced effluent into the Omai River, killing two hundred fish. After five days of silence, Cambior reported the spill, calling it the result of "a completely unusual set of circumstances coupled with human error." But that wasn't the end of the problem.

In August, a retaining wall surrounding the tailings dam cracked, releasing another two thousand pounds (906 kg) of cyanide into the Omai. Hundreds more dead fish were spotted, and there were reports of dead livestock floating down the river. Residents complained that the water smelled bad. After making a nation-wide apology, calling the spill a "very serious industrial accident," Cambior and its partner, Golden Star, dismissed suggestions of any long-term threat to people or the environment. The Guyanese government was not mollified. Guyana's president, Cheddi Jagan, appeared on national television, describing the spill as a "major disaster" and a "national emergency." He added that "although the consequences can be potentially dangerous, I also appeal to you to be calm, and not to panic."

The Guyanese ordered Cambior to immediately suspend operations at the mine, a move that reportedly left company officials fuming. According to environmentalists, Cambior lobbied Canada's High Commissioner in Guyana to try to stop the closure, but the mine was shut for six months, shaving $15 million from Cambior's balance sheet. The Associated Press wire service reported that officials from the mine attempted to mitigate future losses by providing approximately $6,000 to a group of local fishermen in exchange for their promise not to sue Cambior for any loss of livelihood. Cambior says it settled another 226 claims filed against the company at an average cost of $400. Most recently, it has sought a court injunction barring Canadian activists "or any other individual or corporation" from "attempt[ing] to persuade [financial institutions] not to conduct business with Cambior."

Montreal-based lawyer Steve Michelin says that Cambior "has consistently treated this disaster as a public relations irritant." Michelin heads a $69-million civil suit naming three Guyanese plaintiffs. The suit, which was launched in Canada, claims that the tailings dam at the site, "as designed and constructed, was bound to fail." It alleges that, in an effort to slash construction costs by $50 million, Cambior eliminated crucial engineering components designed to protect the dam. Michelin and his clients want Cambior to pay for an environmental clean-up and compensate 23,000 people for damages to property, livestock and livelihoods.

Although a Guyanese commission of inquiry concluded that "faulty construction" caused the accident, it did not hold Cambior criminally responsible, nor did it ban any future discharge of effluent into the Omai River. After repairing the dam, Cambior has its mine running again.

Bob Parsons is not easily shocked. As the former head of Price Waterhouse's world mining group, he's seen plenty of environmental mishaps in the industry. Frankly, they bore him. Parsons says that environmental concerns keep hogging the agenda at major international mining conferences. "I believe social issues are just as important," says Parsons. "I'd rather talk about corruption."

Parsons is in an excellent position to do just that, seeing as he now monitors the mining industry from Price Waterhouse's Indonesian office. Unfortunately, he says, people don't want to hear about the problem. "When lenders like the World Bank make loans to help a country develop its mineral potential, it insists on certain environmental standards," he says. "So why not do the same thing with corruption? Lenders could use their influence to discourage certain behaviours which are not acceptable. Corporations say, 'Well, corruption is a government issue.' What they forget is that development is usually slow in places where corruption runs wild.

It's in the companies' own interests to deal with the problem, and not just play along. People who play with fire get burned."

The issue grabbed Parson's interest several years ago, when he was working on a number of mining projects in Kazakhstan. "I saw a new nation with tremendous potential. The level of education was good. The infrastructure was there. And nothing was happening. People were getting poorer. The elderly were suffering. You could see it on the streets." The reason, he says, was simple. Inbred corruption was discouraging investment from the West. "There's the president, the former head of the local Communist party branch, who said he was going to change everything. It was bullshit. He's now the eighteenth richest man in the world. He and his cronies just grabbed everything in sight."

Now they're selling. As some Canadian mining companies have discovered, doing business in Kazakhstan means currying favour with the right people and making huge payments to "reserve" a deposit. Even that's no guarantee that a project will proceed. Placer Dome Inc. found out the hard way. The Vancouver-based company is no piker. It's a true multinational mining giant, with producing gold mines in Canada, the United States, Australia, Chile, the Philippines and Papua New Guinea, and annual revenues of $1 billion. In 1994, it decided to get involved in Kazakhstan, offering a $35-million "deposit" on an abandoned gold property. After examining the site, Placer changed its mind and asked for the money back. The Kazakhstanis refused and demanded that Placer pay another $270 million, which was the estimated cost of putting the property into production. A Placer spokesman said the company was "ripped off" and suspected that the government spent the original $35 million.

Placer was naive, says Parsons. But other Canadian companies have experienced similar problems in the former Soviet republic. In October 1996, World Wide Minerals Ltd. raised US$25 million after announcing plans to purchase two uranium mines in northern Kazakhstan. "We have done this during a period that has seen... euphoria in the capital markets where virtually anything could be financed," noted a company spokeswoman. Less than a year later, the government refused to grant World Wide an export licence to ship uranium to the United States. The company, which claims to have spent US$23 million upgrading the mines, was forced to suspend operations. The government responded by cancelling World Wide's five-year operating contract, which was "tantamount to expropriation," says World Wide chairman Paul Carroll.

A third Canadian company, Kazakhstan Goldfields Corp., invested US$18.5 million upgrading three abandoned gold mines. By September 1996, it had the mines operating, but when a local smelting plant went bankrupt, the company was forced to close two of them until the problem was solved. Then the price of gold dropped, and Kazakhstan Goldfields temporarily closed the third mine. The government's Department for State Property swooped in, terminating the company's management contracts and grabbing control of the three mines. Payments to certain high-ranking officials proved useless, says a Kazakhstan Goldfields spokesman. "We covered all the bases, and it went for nothing," he says. "The Kazakhstani government got what it wanted, and then ran us off. I don't think the situation will ever improve."

Canadian miners operating in Africa have run into similar problems. Constant political upheaval has plagued efforts to develop mineral properties in countries such as Sierra Leone and the former Zaire, now known as the Democratic Republic of Congo. In their efforts to nail down security of tenure, some companies have cosied up to militant rebel forces bent on overthrowing existing regimes. Diamond Fields co-founder Jean-Raymond Boulle, a Mauritian businessman, signed a provisional contract with Laurent Kabila, a former Marxist turned free-market buccaneer, and leader of the rebel squad that overthrew the Mobutu government in 1997. According to Karen Howlett and Madelaine Drohan, an excellent investigative duo with *The Globe and Mail*, Boulle gave Kabila's rebels the use of a Lear jet and advanced them a $1-million loan on behalf of America Mineral Fields Inc., a VSE-listed company. The move was made as part of an attempt to get the inside track on

US$1 billion worth of deals in Congo, including the large Kolwezi copper project.

Boulle's generosity pales in comparison to that of another Canadian company, Tenke Mining Corp. Howlett and Drohan discovered that the junior explorer had advanced Kabila US$50 million for the rights to develop another copper deposit. In a story published in May 1997, just days before Kabila's coup, Howlett and Drohan reported that rebels indicated the money "would probably finance the war effort." According to the group's "finance minister," Mawampanga Mwana Nanga, the payment would come in handy. "Even if Tenke does not help us directly to capture [Congo capital] Kinshasa, the money gives us room to manoeuvre," he was quoted. Tenke president Ted Webb, who met with the rebel leader a few weeks prior to the coup, said he had "a good thing going with Kabila."

That may be his bottom line, but it provides no assurance that Tenke will hang on to its property. Nine months into the new Kabila regime, Congo was still the same calamitous country, beset with poverty, inflation and a crippled infrastructure. Corruption within the nation's bureaucracy remained rampant. Political parties are banned, and journalists have been arrested. It is a place where the normal rules of law — which North Americans sometimes take for granted — simply don't apply. Early in 1998, Kabila cancelled twelve mining contracts he signed since taking power. The powerful national mining company, Gecamines, said it planned to "re-examine" the Tenke deal. Meanwhile, the arrangement Boulle had negotiated for America Mineral Fields was torn up; according to the Congolese minister of mines, the Kolwezi deal was "fraught with irregularities and illegal points."

Junior mining companies that climb into bed with shaky foreign regimes rarely create wealth; indeed, when the exchange of money precedes due diligence, and secret deals are the order of business, they are accidents waiting to happen. But working abroad, far from prying eyes, Canadian mining companies seem to feel safe playing fast and loose with the facts, embellishing results and doctoring their claims. Analysts and investors who take any of their information at face value, without looking below the surface, are living precariously. The Bre-X debacle is the ultimate proof of that.

9

BENEATH
THE SURFACE

Busang hit a 'Golden Pot' at hole 3, 4!!
– Internal Bre-X memo from Michael de Guzman,
December 1993

We don't want to do anything that's not legal at Bre-X.
We want to adhere to the mining laws.
– John Felderhof

"**THE TRUTH?**" The man with the trim white beard rocked on his feet and smiled. "The truth is there's a mining boom going on. It's coming from Canada, and this is where it's heading." He pointed to a map of Borneo. "Kalimantan. It's not hype, either. This one is real."

He believed it. I believed him. This wasn't some sleazy shill with stale beer on his breath. Kurt MacLeod (not his real name) was dead convincing. An elegant, likable man, he had nearly forty years in the mining industry, ten of them in Indonesia. He knew the place inside out. The boardroom where we sat was a veritable shrine to Indonesia, decked out with the red and white national flag and framed photographs of President Suharto and then-Vice-President Try Sutrisno.

A large, psychedelic statue of Garuda, the mythological bird that Indonesians revere, sat proudly on a bookcase, poised as though ready for take-off.

We met a few times at his set of offices on Bay Street, back when Busang was still real. MacLeod joined the Indonesian gold rush in late 1995, piecing together six projects, four of them in Kalimantan, the largest right beside Busang. He had another four diamond prospects elsewhere in the country. His company was just starting its exploration campaign, and the future looked great. A quick private placement raised $9 million. His stock was moving up, fast. Journalists were paying him visits.

MacLeod struck me as a decent promoter, the kind you'd like to see succeed. He didn't brag, or try to push his stock. He didn't bullshit about Indonesia, either. He laughed when I told him about a visit I'd made to the Indonesian consulate in Toronto. The consul, a formal, slightly intimidating woman named Titiek Suyono, had leaned over her coffee table and told me, very firmly, that "there is no corruption in Indonesia."

The reality is that Indonesia is a hotbed of graft, one of the worst in the world, according to annual business polls. MacLeod had been hit up for bribes plenty of times. It no longer fazed him. "Pay-offs are used to speed up the system," he explained. "North Americans have this holier-than-thou attitude when it comes to corruption. But they forget that it takes two. Whenever there's a taker, there has to be a giver. And we give just as much as anyone."

Paul Jaurie walks around Jakarta with a large white envelope, stuffed with crisp fifty-thousand-rupiah banknotes, which used to be worth $30 before the Indonesian economy tanked. A mining industry "facilitator" with ties to several Canadian companies, he smooths out the inevitable hassles that arise when dealing with Indonesia's civil service. An Indonesian of Chinese descent, he deals mainly with low- to mid-level functionaries who can't make ends meet on their puny, $100-a-month salaries.

While he insists he leaves the high-level graft to other people, Jaurie does spend a lot of time inside the Department of Mines, a bleak, dimly lit building on a busy Jakarta thoroughfare. He knows most of the senior civil servants by name. "I do favours for them," he says, "and they do favours for me. Some foreigners think they know everything," he says. "But here they are blind. They don't know who to pay, or how much."

Most of the time, it's petty stuff, a few hundred dollars to secure a "missing" document, maybe an airplane ticket and a hotel room for a travelling bureaucrat. Up at the top, at the cabinet level, well, that's another matter. "My God, it's disgusting the amount of money the bastards are taking," an official with a large Canadian mining company told me. "It's not like the old days, when $20,000 fixed everything."

Indonesia's top mining technocrat was walking on air. Kuntoro Mangkusubroto thought Bre-X was a godsend. The company first caught his attention in 1995, when its resource estimates started to skyrocket. With Bre-X pouring it on in Busang, dozens of delegates from other Canadian companies were arriving in Jakarta to negotiate land titles. Kuntoro was the man everyone wanted to see. "I'm the one who promoted Indonesia," he told me, during one of several "post–Bre-X chats" we shared in Jakarta. "In that promotion, there was Busang, which eventually became a scam. I'm responsible. I cannot avoid it."

Energetic and bright, with a youthful, Javanese face, Kuntoro had made a name for himself as a reform-minded, free-market specialist. Like a lot of Indonesia's elite, he was educated at the Bandung Institute of Technology, a prestigious institution outside Jakarta. He went on to study at Boston University, and then obtained a doctorate in engineering from Stanford University. In the late 1980s, he was credited with turning an inefficient, state-owned tin company into one of Indonesia's most profitable private enterprises, laying off twenty thousand employees in the process.

Kuntoro was richly rewarded. He was named Director General of the Department of Mines, giving him direct access to political figures and the heads of major multinational corporations. According to Kurt MacLeod, Kuntoro wasn't above accepting small favours. His monthly wage of $500 — excellent by Indonesian standards — was supplemented by regular payments from a special *yayasan*, a "charitable foundation" that the mines department had set up. Kuntoro sat on its board. The *yayasan* collected money from foreign mining companies, in the form of unspecified donations. Kuntoro had his own private secretary, a car and a driver, and access to a country club.

While in office, Kuntoro implemented a series of Western-oriented reforms that gave foreign mining companies more control over their operations inside Indonesia. He phased out domestic ownership requirements and reduced taxes, in an effort to attract more investment. While the changes were unpopular inside nationalist circles, the international community applauded Kuntoro for his open-minded approach. He was fond of displaying a study conducted by an Australian bank that ranked Indonesia as the most attractive Asian country for mining investment, based on local laws, political risk and geological prospectivity. Money was flowing into his department's coffers. When the junior explorers began knocking on his door, applying for Contracts of Work, they had to lay out a $5-a-hectare "seriousness bond" to secure the property they wanted. In two years, the mining department's bank account swelled to US$350 million.

Kuntoro didn't rise to the top of the department without a Machiavellian touch. He was obsessed with gaining President Suharto's favour. A good performance was bound to improve his position inside Indonesia's labyrinthine bureaucracy. He told me he hoped to be named rector of the Bandung Institute, an appointment that required Suharto's approval. There were also whispers he was lobbying to become minister of mines. The Bre-X discovery was going to legitimize his efforts to modernize the industry, or so he believed. But rumours eventually surfaced that Kuntoro was in Bre-X's pocket. One story had him accepting a house in Perth, Australia, as payment for certain services rendered. Kuntoro isn't offended. "I have heard the talk," he calmly admitted to me, in May 1997. "I was in collusion with Bre-X, I had stock in the company, I have a house in Malibu. It's all nonsense."

But in his haste to boost Indonesia and enhance his own reputation, Kuntoro allowed Bre-X to bend all kinds of rules. He knew the company was conducting its massive exploration campaign in the touted Southeast Zone without proper authorization. Bre-X was entitled to do some minor surface work on the property, using hammers and hand drills, but heavy drilling was expressly prohibited. Bre-X had only a preliminary investigation permit, known as a SIPP, yet it was acting as if it had already obtained a coveted Contract of Work.

Legally, Bre-X had no right to drill in the Southeast Zone, but Kuntoro made a conscious decision to look the other way. "Maybe Bre-X was going too far," he said, shrugging. "But I felt there needed to be some flexibility. When things are done more quickly, it's the exploration company that assumes the risk, not the bureaucracy. It's not our money that is being spent." Other technocrats inside the Department of Mines took the same position. "It's not a serious mistake," Rozik Soetjipto, Kuntoro's second-in-command, told me, months before Busang was exposed as a hoax. "In my mind, it's good. Things move faster. We can give them mercy."

In February 1996, Kuntoro was told that Bre-X was about to boost its gold estimates, to somewhere between thirty million and forty million ounces. Walsh and Felderhof had already hinted at the figures in a press release, but the news had not been picked up by the Indonesian journalists. Few Indonesians outside of the mining department had even heard of Bre-X. Kuntoro, to his everlasting regret, quickly changed all that.

"I was in a state of euphoria," Kuntoro recalls. It was Ramadan, the Islamic fasting period. He was hosting a reception in his office one evening and started chatting to a local reporter. Kuntoro mentioned that Bre-X "had found" forty million ounces of gold. "It wasn't an official announcement," Kuntoro says. "It was the end of the day and we were breaking our fast. It was a very informal setting.

But the news went out." Kuntoro saw no reason to tell anyone the forty-million-ounce figure was an unconfirmed estimate. "Any finding, I was glad to put out in the public," he says. As it turns out, the information was wrong. Verification, he says, "was not my function."

Kuntoro never visited Busang. The property was closed to everyone but the pros who covered Bre-X for the Street. People who might have been expected to dig deeper were kept away. The analysts were no threat; they accepted everything at face value and returned home with identical praise. *Geological evidence is very strong... a very professional exploration campaign... easily verified... among the largest gold deposits in the world.* Geologists with other Canadian-based companies frequently asked to see the miraculous Busang operation. They were always rebuffed. "It was very unusual for a Canadian company to refuse visits," says South Pacific's Cam DeLong. "The normal practice is to offer exchange trips. You show them your stuff, and they show you theirs." That should have worried the technocrats in Jakarta, and the analysts on Bay Street, but they never mentioned Bre-X's strange secrecy.

Then again, the Busang scam was an artful dodge, directed at ground level with precision and skill by Bre-X's field commander, Michael de Guzman. He ran his clandestine activities for three years, while Walsh and Co. broadcast the illusion to the public. He seemed content to let Walsh and Felderhof feed their own egos in front of reporters and photographers, while he crept in the shadows. He didn't complain when Felderhof took most of the credit for finding Busang. De Guzman projected himself as a happy-go-lucky fellow, all smiles and chuckles and sympathetic shrugs.

Everyone liked Mike. Aji Maulana remembers meeting him in 1986, when the geologist made his first trip to Busang. The owner of a transportation company, Aji took him upriver in his speedboat, staying in Long Tesak for a couple of days while his client prowled the jungle. Six years later, in 1992, de Guzman made another trip to Busang and came back with his glowing report for Willy McLucas.

He returned the following spring, this time with three Indonesian geologists in tow. They did some surface sampling at the time, but their primary job was to build up the camp left behind by the Australians. De Guzman and the Indonesians spent two weeks at a time at Busang, and then returned to Samarinda to shop for supplies. Two or three days later, they would head back to the property. This went on for two months.

By the fall of 1993, more men had begun to arrive — truck drivers, rig-pigs, cooks, cleaners. De Guzman surrounded himself with a cadre of Filipino geologists, most of whom he'd met while working for Benguet Corp. in the Philippines. There were Bobby Ramirez, John Salamat, and Sonny Imperial, all from Benguet. Rudy Vega and Manny Puspos were later added to the crew. De Guzman's chief assistant and closest friend was Manny's brother, Cesar Puspos, a shy, retiring thirty-year-old who became Busang's senior project manager. Finally, there was the despised site manager, Jerry Alo, forty-eight, who looked after day-to-day affairs. The Indonesians hated Alo. "Jerry was a bad man," says Cornelius, an electrician who worked at Busang for three months. "He was supposed to pay us, and he was always late. People were suspicious of him. We thought he was stealing our money."

There were only a handful more expatriates at Busang. Most were young Australian and Canadian geologists who either performed mundane tasks such as logging core samples or were sent to remote fly camps to grab surface samples and conduct geochemical surveys. Trevor Cavicchi, a geologist from Nova Scotia, hired by Bre-X in June 1996, says the Filipinos basically kept to themselves, huddling together at mealtime and maintaining a rather superior attitude to the rest of the men. What little interaction they did have with their co-workers was usually confined to a sweaty game of basketball outside the mess hall. Sometimes, they crossed paths at the old prostitution complex down the road.

The Filipinos did not seem terribly qualified to run what would eventually be billed as the biggest gold deposit on the planet. Puspos and Alo were obviously overwhelmed by their management duties and frequently had to call upon de Guzman for advice. The one man who seemed best suited to run the camp, Jonathan Nassey, was shunted off to one of Bre-X's other, far-flung properties. Nassey was more familiar with Busang than just about anyone, perhaps even de Guzman. He had done some of the early exploration work there; in fact, he'd been exploration manager for PT WAM, the joint-venture company Bre-X eventually acquired. He was a colourful character who had once run for governor of his native province of Irian Jaya. Peter Howe, who describes Nassey as "a hell of a nice guy," hired him to lead one of his exploration companies in the 1980s. Felderhof was fond of Nassey too. The two men had set up an eco-tourist operation together in Irian Jaya after the Australian-led exploration boom died down. Yet Felderhof said nothing when de Guzman ordered Nassey off Busang.

Felderhof hired PT Drillinti Tiko to conduct the first diamond drill program in the Central Zone. The drillers moved up and down the site, punching holes every fifty metres along a mineralized fault, beside the area drilled years earlier by the Australians and then down into the Southeast Zone. Eventually, Drillinto Tiko got the contract for the entire operation. Rumour had it that de Guzman got a healthy kickback from the company, about $10 for every metre drilled. More than ninety thousand metres of raw core was pulled from the Central and Southeast zones between 1994 and 1997.

There were other stories about de Guzman, incredible tales about his sexual escapades. Some turned out to be true; de Guzman was an incorrigible womanizer and had a handful of wives scattered strategically about Indonesia. In addition to Teresa, who had remained in the Philippines with their six children, there was Susani in Sulawesi, Lilis in Samarinda, and another unfortunate creature named Gini, who lived just outside of Jakarta with their son, Big Boy. Apparently, none of them knew their husband was an avid polygamist. De Guzman lavished money and sickly sweet little love notes upon all his wives. Somehow, he managed to keep all of them in style, despite his modest salary.

De Guzman's fourth bride, a simple, beautiful twenty-two-year-old from Samarinda, was swept off her feet by the fun-loving man who crooned songs in her ear. The poor girl had no idea. She was easily seduced with flowers and candies. De Guzman got her a secretarial job at the Bre-X office in Samarinda, married her two years later in a Catholic church and installed her inside a sparkling new $200,000 townhouse on the outskirts of town. I drove by the place one day on a borrowed scooter. Lilis' new Suzuki sport utility van sat in the driveway. Her father stood frowning on the patio out front, arms crossed, zealously guarding the spread from nosy intruders.

De Guzman had other secrets. He was sick with hepatitis B, according to his third wife, Susani Mawengkang. A Muslim woman, Susani married de Guzman in 1995, after he convinced her he had converted to Islam and changed his name to Ismail Daud. They had one child together, a boy named Anthony. Susani recalls that de Guzman was frequently racked with pain. "He would complain of kidney or liver pains and break out in a deep sweat all over his body," Susani says. "He would cry often with me." De Guzman sought all kinds of treatment for the disease, including folk medicine, with little success. A visit to a *dukan* — or traditional healer — in Kalimantan ended in disaster, after an unusual treatment left ugly lesions on his back.

His personal life may have been chaotic, but de Guzman was an able field marshal. Once in full swing, the Busang operation was extremely efficient. Bre-X spent over $20 million at Busang, and it showed. The original camp took on an air of permanence; a dozen solid-looking buildings dotted the site, connected by tidy gravel pathways. A fully serviced executive suite perched on a small rise, overlooking a scenic pond. The grounds were nicely manicured; a flower bed spelled the magic words Bre-X Minerals on the side of a hill. Closer to the Southeast Zone, a second camp housed senior-level workers. Then there was the new town site, with its eighty identical houses and a mosque.

As the amazing results began to roll in, three more diamond drill rigs appeared, and truckloads of men. Bre-X even decided to buy forty-nine per cent of Drillinti Tiko for US$1.5 million. David Walsh

personally advanced the money to Drillinti Tiko's private holding
company, Far East Mining Services Ltd. in early 1996. The deal was
never finalized, say officials with Drillinti Tiko, because "it was
deemed undesirable to have our major client seen by the mining
industry as a part shareholder in our company." However, the
money was never returned and ended up on Bresea's books as an
outstanding "loan."

Busang had become a serious exploration exercise. The focus had
completely shifted from the Central Zone to the Southeast Zone,
where Felderhof, de Guzman and Cesar Puspos claimed to have
identified a cigar-shaped diatreme dome, twelve kilometres long,
and six kilometres wide. Visible from the air, the protrusion never
garnered any interest until a bulldozer uncovered what they claimed
was a tell-tale rock aberration. This, they decided, was the mother
lode.

"The Central Zone pales in comparison to the potential of the
Southeast Zone," Felderhof said. "The Busang project in its entirety
has the potential of becoming one of the world's great gold ore
bodies." Proving up thirty million ounces would be a snap, he said.
Bre-X could delineate that kind of resource in a matter of months.

The emphasis was always on speed. Hundreds of metres of heavy,
cylindrical core were pulled from the ground every day, washed
down and placed in wooden trays. It was trucked into camp and
inspected by waiting geologists. "Good" core, the stuff that was supposed to contain the proper mineral qualities to support gold, was
culled from the "bad," cut into one-metre sections, and numbered.

Rather than follow industry practice and save half the core, de
Guzman and Felderhof sent it all to the laboratory, minus a ten-
centimetre piece, the so-called "skeleton core." This was sliced from
each mineralized section and put into storage. The remainder was
put into clear, plastic sample bags, labelled and trucked down to the
wharf at Long Tesak. From there, the bags were placed on a pair of
speedboats, which ploughed down river to a wharf at Samarinda.
Then it was trucked to Indo Assay in Balikpapan, where it was
assayed. At least, that was the story Bre-X put forward. According to

a flow chart that the company circulated, the core was immediately trucked to the lab. But there was a crucial step in between that Bre-X neglected to mention.

I met Aji Maulana at a Samarinda hotel in May 1997. A small, restless man, Aji had just come from an interview with the local police. They wanted to know more about the work he had done for Bre-X. I did too. Aji probably knew the Bre-X characters better than anyone in Samarinda. He'd worked with them just about every day for more than three years, delivering their core samples to the assay lab. He figured he'd been to Busang a thousand times in his speedboat. At $450 a trip, Bre-X had been his bread and butter.

We sat in a corner of the hotel lobby. Aji couldn't sit still. He giggled nervously and fidgeted with my name card, flipping it over and over, tapping it with his finger. After a few minutes of pointless conversation, he finally opened up. His job had been pretty straightforward, he said. Every day, he and his kid brother or another employee drove a couple of speedboats up the river to Long Tesak, where they picked up twenty to thirty bags of core. They proceeded down to Samarinda and loaded the bags on to trucks, just as Bre-X had said. But the bags didn't go straight to Balikpapan. Instead, they took the core to a warehouse in Samarinda, where, Aji said, it sat for three or four days. "Rudy and Jerry were there," he added. What they were doing, he claimed he didn't know.

I told Aji that this didn't jibe with Bre-X's version of events. Bre-X never mentioned a layover in Samarinda. He suddenly became defensive. "I have no interest in Bre-X," he told me. I asked him what he meant. He shook his head. "I am sure Bre-X did not mix gold with the samples." His comment surprised me, since the subject of tampering hadn't been broached. Then he dropped another clanger. "I have seen the gold," he said. "Specks of gold, all up and down the core." This echoed de Guzman's claim, that he had seen gold in the core samples.

They both lied. The gold didn't exist. Three weeks earlier, Strathcona Mineral Services Ltd., the Toronto engineering firm, had announced that Busang was a fraud, that Bre-X had salted its core, most likely at the Samarinda warehouse. Someone had opened the bags inside the heavily guarded facility, dropped in a few tiny grains of gold and then had them resealed. Performed carefully, the procedure would ensure that each core sample showed a consistent grade. It took only a few hundred grams of gold dust to do the entire job.

Strathcona president Graham Farquharson says the salting likely began in the Central Zone, as early as 1993, when the first drill core came in from Indo Assay. Tests conducted by Lakefield Research Limited revealed that the gold "found" in holes three and four was actually man-made gold-copper alloy commonly found in jewellery. As the scheme progressed and the salting techniques improved, the perpetrators apparently began to rely on finer, alluvial gold, which is by all accounts easily obtainable from local Dayak panners. De Guzman allegedly had his underlings buy gold from the Dayak, telling them he needed it to fashion rings for his wives and children.

Expatriates working at Busang had concerns about the sampling process, although they remained quiet until the scam was exposed. Four Canadian geologists told investigators that they had never seen any gold at the site; they had, however, observed bags of core sitting open inside a billiards room at the Samarinda warehouse during 1996 and 1997. The caretaking staff at the warehouse noticed the same thing and said that Cesar Puspos and de Guzman often stayed late inside the billiards room, doors locked. One young geologist claims he asked Cesar Puspos why the bags had been opened. Puspos said that he was "checking to make sure poor-quality rock did not go into the bag." None of the Canadians say they actually saw anyone drop gold inside the bags, but the very fact they had been opened violated accepted industry practice, exposing the rock to all kinds of elements and temptations.

The tampered core was then shipped to Indo Assay in Balikpapan, where the bags were opened again. The contents — including any fragments lying at the bottom — were shaken into a crusher and ground to a powder. At this point there were sizeable bits of gold floating around, some almost half a millimetre in diameter, with a circumference similar in size to the tip of a pencil. The gold was leached out using a cyanide solution, and that's how the fabulous grades were determined.

It made perfect sense. For one thing, it explained why Bre-X insisted on crushing the entire core, instead of splitting it down the middle. Crushing left no means of double-checking the lab's results. The only way to confirm the grades would be to twin the holes back at Busang, something Bre-X had steadfastly refused to do, citing time and money constraints. Strathcona's conclusions also explained why the skeleton core that Bill Stanley and the other analysts had brought home tested negative. Their samples were virgin bedrock, untouched by de Guzman and the others inside the Samarinda salting mill. The stuff Indo Assay received was already tainted.

Back in Canada, the popular spin was that the Filipinos had done it, no one else knew, end of story. But Aji's comments bothered me. It seemed he knew more about the mysterious Samarinda stopover than he wanted to let on. The very fact that Aji and his men were aware that Bre-X was moving its core through the warehouse was significant. If they knew, how could it have possibly escaped the attention of John Felderhof and David Walsh?

At some point, the Bre-X executives must have wondered where all that visible gold was coming from. A person could have stared at the Busang core for hours with a magnifying glass and not seen any yellow metal. What's more, by 1996, Bre-X and its vaunted engineering firm, Kilborn, had information that clearly suggested that something wasn't right. Metallurgical tests Kilborn commissioned made note of some anomalous-looking gold grains found in the crushed Central Zone core. Most of the gold particles were large, loose and worn down from erosion, indicating they likely came from an alluvial source such as a river. Gold grains that come from a hard-rock — or primary — source such as Busang are normally microscopic in size, wiry in shape, and bound to other bits of material.

Normet Pty Ltd., the Australian company Kilborn hired to examine Busang's mineral content, never came out and said the drill core had been dusted with gold, but it was one obvious explanation for these curious results. Instead, Normet recommended that Bre-X drill parallel holes at Busang, to confirm the presence of gold. This was ignored, something that worried even Paul Kavanagh, as he later admitted to investigators. "[Normet personnel were] engaged to do their job," he said, "and if they feel in doing their job they want to do that, then I would find it difficult to see why Bre-X couldn't have spared one of its drills for an outfit to drill a hole."

Another company that was hired to examine the core sampling procedures criticized Bre-X for not splitting its core. Mineral Resources Development Inc., based in San Mateo, California, urged Bre-X to save half its core for independent verification. Once again, Bre-X ignored the suggestion. In fact, the company buried all its troubling technical reports, never sharing them with the public. In the meantime, de Guzman and his friends kept shaking the salt. As long as the reserve estimates kept moving the stock, everyone was happy.

10

MUNK'S RULES

Life is about meeting objectives. Sometimes your objectives cross other people's. Then you have to fight — and you fight to win.

If an acquisition is strategically right, don't worry about the price.
   – Two of Peter Munk's "golden rules"

BARRICK GOLD had not forgotten about Bre-X. The company was still smarting from its embarrassing dalliance with David Walsh two years earlier. Its failed attempt to purchase a majority stake in Busang had enraged Barrick's chairman, Peter Munk, who was used to getting what he wanted. He came to covet Busang and was determined not to let the alleged deposit slip through his fingers again. It didn't matter that Walsh refused to talk to him. He didn't need his cooperation. Munk was prepared to play hardball, even if it meant risking his company's reputation.

Brilliant or ruthless — take your pick. Munk was no shrinking violet. "He's a very impatient person," says one of his staff members. "He believes in instant gratification." Born in Hungary in 1927, Munk grew up in a wealthy Jewish household. When the German army marched into Budapest in 1944, his family fled to Switzerland, paying for their freedom with gold coins. After completing high school in Zurich, Munk shipped out on his own, landing in Canada. Life wasn't easy for the frail-looking immigrant. Jobs were hard to come by, but Munk survived his first summer picking tobacco in southern Ontario. He saved enough money to enrol in the University of Toronto and spent the next four years studying electrical engineering. A salesman at heart, Munk fed himself by flogging Christmas trees every winter and installing expensive hi-fi equipment in private homes. He got to know his customers, catering to their fancy tastes and emulating the trappings of their success with expensive clothes and affected mannerisms. Munk was not going to settle for a second-rate lifestyle. He was going to make a fortune and look good doing it.

Appearances have always meant plenty to Munk. His first real business venture, Clairtone Sound Corp., made expensive hi-fi equipment, encased in beautiful wooden cabinets. From the outset, Munk tried to make Clairtone synonymous with style, putting "class" ahead of performance. He wanted goods that appealed more to the eyes than the ears. The equipment always looked great; in the 1960s, plastic casing replaced wood, and the hi-fi's assumed a curving, modernist form that won all kinds of design awards. Putting form ahead of function may have worked in the short term, but the strategy was bound to fail eventually. Clairtone never showed any real profits and always carried significant debt. Munk managed to squeeze millions of dollars in subsidies from the government of Nova Scotia, which built the company a new manufacturing plant in a small mining town, but Clairtone went belly up in 1971, sticking taxpayers with a $23-million tab. By then, Munk was long gone.

Real estate. That was the next ticket. Munk developed land in Fiji and bought a fifty per cent share in Travelodge Hotels, which owned a string of inns across the South Pacific. This looked like another bust; Munk spent far too lavishly on the publicly traded Travelodge. Its management was incompetent, and Munk lacked influence at the board level to change it. He watched helplessly as his shares quickly slid from $1.55 to 24 cents and thought about walking away from the investment. He launched a counter-attack instead, after hooking up with Adnan Khashoggi, the notorious Saudi dealer of cheap American automobiles and deadly military hardware. Munk met Khashoggi in the mid-1970s, when both men were attempting to develop luxury resorts in Egypt. Munk used Khashoggi's financial might to buy a majority stake in Travelodge, then making strategic alliances with key directors and orchestrating the removal of its founding chairman. Travelodge was privatized, rationalized and re-energized, and then sold to a Malaysian businessman for $130 million.

Munk promptly invested his windfall in the oil and gas business. Given the timing, this was a big mistake. Petroleum prices had peaked when Munk bought his first producing outfit in 1981, for $60 million. He dropped $30 million on a bunch of dud properties, and so ended that little adventure. Once again, Munk's passion had interfered with his reason. "We are not exactly experts when it comes to talking about drilling wells and the rest of it," he told his sympathetic biographer, Donald Rumball. It was a costly error in judgment, one that Munk was destined to repeat more than a decade later.

Munk risked more than money when he went after Busang for the second time; he played an audacious political game that eventually blew up in his face. Barrick Gold, the company Munk built from scratch, became an object of scorn inside the mining industry for the underhanded way it conducted itself in Indonesia.

Barrick didn't need Busang; the company was already one of the top three gold producers in the world, with an annual cash flow of $500 million. First listed on the Toronto Stock Exchange in 1983, Barrick began with a couple of tired old gold mines in Alaska. Three years later Barrick was listed on the Paris Bourse and made its first key acquisition, the twenty-five million-ounce Goldstrike deposit in Nevada, for a bargain sum of US$62 million. Using Goldstrike as its foundation and cash cow, Barrick reached out, snapping up whole companies, including Lac Minerals Ltd., for a staggering

US$1.7 billion. By 1996, Barrick was among the most productive gold companies on the planet.

But Munk wasn't content. He wanted to lead *the* biggest mining company. That meant venturing outside the western hemisphere and into unfamiliar territory blossoming under global trade. To help achieve this objective, Munk had already recruited a couple of former political heavyweights to lead his international advisory board. George Bush and Brian Mulroney, old fishing buddies and inveterate free-trading schmoozers, were hired to help give Barrick access inside foreign regimes. They didn't come cheap. Mulroney, who was also a Barrick director, was paid $140,000 in 1994, for "advice and expenses." Canada's former prime minister was also handed 500,000 share options.

In 1995, Barrick opened a small office at the Cilandak Estates in Jakarta, filling it with an experienced administrative staff. The company then signed a series of option agreements with an American company, Yamana Resources Inc., which controlled twelve Contracts of Work in East Kalimantan. These covered almost three million hectares of property immediately north of Busang. If Yamana's geologists turned up any viable gold reserves, Barrick reserved the right to build and operate a mine and take seventy-five percent of the output. It was a reasonable arrangement that cost Barrick next to nothing. More important, however, it gave the company a beachhead, from which Munk planned to grab hold of Busang.

Munk and his staff devised a number of interesting strategies. He initially relied upon the experience of his new commercial manager, Tim Scott. The amiable Australian had lived in Indonesia for more than thirty years and knew the country inside-out. A former intelligence officer with the Australian secret service, he'd witnessed the 1965 coup and later developed close ties with many of the country's top political and military figures. In fact, he had translated into English the memoirs of Indonesia's army commander, Benny Murdani, a key Suharto loyalist who ended up as the patronage-dispensing secretary of state.

According to a source within Barrick who has an intimate knowledge of the events, Scott "tried to open a back door to Busang." He counselled Barrick to quietly strike a deal with Bre-X's Indonesian partner, a respected timber magnate named Haji Syakerani. Through his company PT Askatindo, Syakerani held a ten per cent interest in the Central Zone and another ten per cent on the COW application in the Southeast Zone. The idea was for Barrick to buy out Syakerani and then take Bre-X in a hostile takeover.

In late 1995, Munk dispatched a pair of lawyers to Jakarta to meet with Scott. The three men penned a draft offer for Syakerani and, without thinking, left it sitting on top of a desk. An Indonesian geologist employed by Barrick, got his hands on the document and showed it to another Canadian company, Minorca Resources Inc. By the time Scott got around to meeting with Syakerani in person, it was too late. Minorca announced it had signed a deal with Syakerani, giving it control of his stake in the Southeast Zone. "Some of the wording of their agreement was a straight lift from our pen," says the Barrick source.

It was a setback, but Munk could at least take comfort that the plot hadn't been exposed. By February 1996, press coverage of Bre-X had exploded. Every business journalist in North America had tuned into Bre-X, but they were all focused on the company's amazing market run, treating it like some endless lottery story, and were completely unaware of the serious wrangling going on behind the scenes.

Bre-X shares were trading at $170, giving the company a market value of $4 billion. Walsh was now worth $300 million on paper. He'd already cashed in 178,000 shares for more than $9 million and bought a US$1.6-million estate in Nassau, the Bahamas. Few people begrudged him the money. Rumpled and nicotine-stained, he was being hailed as a capitalist hero who'd made millionaires out of plain ordinary folk. He revelled in the role. "It's unbelievable," he said. "People, you know, come up and say we changed their lives, and they can put their kids through university, and they got tears in their eyes."

Some people forgot that Bre-X was still a tiny exploration outfit that couldn't conceivably develop Busang on its own. Walsh was still operating Bre-X from his basement in the suburbs of Calgary. That changed in March, when the company moved into a $2-million office building it bought on the edge of a trendy downtown neighbourhood. He hired a sharp numbers man, Rolando Francisco, and made him chief financial officer. Francisco gave Bre-X some extra credibility. A native of the Philippines who started his accounting career in a pencil factory, he had developed an excellent reputation within the Canadian mining industry and had lots of contacts. Ian Hamilton worked with him at Lac Minerals for ten years. Francisco came on board just after Lac had been hosed in the New Cinch salting scam. "Roly rose from nothing to become a very capable executive," says Hamilton. "He was a tough negotiator. He really knew how to play people off on each other. He could always squeeze a few dollars out of someone."

With Francisco in tow, people assumed Bre-X was preparing to negotiate a major deal. In fact, immediately after assuming his place at Bre-X, Francisco travelled to Busang with the managing director of gold development for CRA Ltd., one of the world's leading mining companies. They were given the royal tour by Felderhof and de Guzman, and some technical information was released to CRA, indicating that the two sides were interested in formulating some kind of a joint venture. Still, Walsh publicly dismissed suggestions that Bre-X was looking for a senior partner, saying he wanted to see how much gold was in the ground first. "We've had continued interest from every major mining company in the world," he said. "That's become disruptive for me to get any work done so I wrote to them all last fall saying it's our intention to increase shareholder value by continuing to drill for our own account."

"Increasing shareholder value" became Walsh's new mantra. He chanted it every time someone asked him when Bre-X would finally make a move towards production at Busang. The gold wasn't going anywhere, he'd say. There'd be time to talk business later. First, he had to "increase shareholder value." He planned to split Bre-X stock ten for one. Walsh had some more private placements in mind and many more meetings with analysts. He was listing his stock on the big boards, in Toronto and Manhattan. The TSE and the NASDAQ were eager to see him. He was very busy raising money, all in the shareholders' interest.

Robert Van Doorn, the analyst from Loewen Ondaatje McCutcheon, was astounded by Walsh's foot-dragging. "We got into a bit of a fight," says Van Doorn. "We kept telling Bre-X to get a senior partner. Walsh said he wanted to drill out the property first, but it was much too big for him. A larger partner would have sped up the exploration."

There was, perhaps, another explanation. Bringing a partner into the picture would finally force Bre-X to account for its reserve estimates. Strangers would show up on the property with their own drilling rigs, expecting to find gold. Once that happened, the party was over.

Barrick was not going to let Walsh control the process. In April, according to private correspondence, Barrick made "preparations for an unsolicited bid to acquire a majority stake in the common stock of Bre-X." But that would be hideously expensive, perhaps $5 billion, which didn't include the heavy capital costs at the mine site. Munk abruptly changed gears a month later, after reading a report prepared by Kroll & Associates, an American investigative firm. Barrick had hired Kroll to check for cracks in Bre-X's corporate armour, and it came back with a binder twenty-five centimetres thick. Kroll noted that Bre-X was technically violating Indonesia's mining law, thanks to its advanced exploration efforts in the Southeast Zone. This was good information. Barrick could use it.

There was more. During a visit to the Securities and Exchange Commission in Washington, D.C., Kroll saw that Bre-X had fudged its application for a NASDAQ listing, by filing a significantly altered copy of Indonesia's standard Contract of Work agreement. It appears the change was made to discourage predators from taking a shot at Bre-X. Section 29 of a COW states that a company may transfer its ownership rights in a property to another company, as long as it has the approval of the minister of mines.

The altered COW that Bre-X filed with the SEC contained a new paragraph, adding that the rights "can only be considered by the Minister *after the Company has reached commercial production.*" (My italics.) Had it been genuine, this caveat would have prevented companies such as Barrick from ever gaining control of Busang. But it looked like a fake, and this was very good information. Kroll also found that Bre-X did not necessarily have clear title to the Central Zone. Bre-X had never told the government about its purchase of PT WAM, the joint-venture company that owned the fourth-generation COW. And as if that weren't enough, one of PT WAM's original Indonesian partners was claiming he had a right to a share in the Southeast Zone. Yusuf Merukh, a controversial businessman and politician, contended that he had been cut out of his rightful piece of the Busang pie. He wanted it back. Jackpot. Barrick's lawyers were ecstatic. "The documentation we reviewed did *not* confirm that Bre-X had legal rights to Busang," noted one member of Barrick's legal counsel. "From Merukh's perspective PT WAM — not Bre-X — was legally entitled to Busang."

From Barrick's perspective, these were all issues worth exploiting. But the company still needed a strong ally inside Indonesia. Bre-X obviously had the support of the technocrats. Cutting some kind of deal with Merukh was a possibility, but he was reputed to be a loose cannon and completely untrustworthy. Barrick would have to sit down with him eventually, but not now. There was one obvious route, thought Peter Munk, who never settles for second best. He would go straight to the top, to Indonesia's first family.

Members of Barrick's Jakarta staff were dismayed when they heard that Munk had decided to enlist the support of Siti (Tutut) Hardiyanti Rukmana, President Suharto's eldest daughter. One employee dashed off a note to Barrick's brass in Toronto, advising against the move. It might suit the company's short-term objectives, he wrote, but in the long run, it was bound to cause headaches.

"I expect the decision came through George Bush," the Barrick employee told me. "He'd been told by Henry Kissinger [a director of Freeport McMoRan Copper & Gold] that the best partnerships were done with the royal family, which was bullshit. You can never control your destiny once you're in bed with the Suharto children. They demand money, and then they want more. They say give me this, and give me that, and it never ends. You can never budget anything, never control anything. You're on the ropes the whole time."

Many Indonesians pine for the simple, carefree days when the president was content to leave business matters to his beloved wife, Siti (Tien) Hartinah. Known throughout the country as Madame Ten Percent, she raised money the old-fashioned way — by holding people up. Her chief bandit was a pistol-packing garrison commander named Eddy Naliprya. Whenever Madame Ten Percent needed some pocket money, she would send Eddy down to the market to do some impromptu "tax collecting."

Things changed in the 1970s and 1980s, as the six Suharto kids reached adulthood. Their doting father set them up with investments in just about every commercial enterprise — banking, agriculture, construction, communications, to name just a few. Son Bambang was handed the exclusive right to import plastic goods, a profitable arrangement upon which he built one of Indonesia's largest conglomerates, the Bimantara Group. Tommy Suharto, the president's youngest boy, was granted the lucrative clove-trading monopoly in 1991. Cloves are a key ingredient in the manufacture of *kreteks*, extremely addictive cigarettes that half the Indonesian population smokes incessantly. Tommy created the country's first private airline and in 1996 was given control of an ill-conceived "national" car program. Tommy's sedans, sluggish gas guzzlers that are manufactured in Korea and imported — duty-free — into Indonesia, have failed to impress anyone, despite their low price. Sales were so slow in the program's first year that Tommy was forced to cut production in half.

Of all the Suharto kids, it is Tutut who wields the most clout, both economic and political. She is one of the richest, most influential figures in Indonesia, and her business conglomerate, PT Citra Lamtoro Gung Persada, has interests in construction, toll roads, broadcasting and electronics worth close to $1 billion. Her husband owns the national Coca-Cola franchise. Tutut also has her father's ear, becoming the president's most trusted advisor after Madame Ten Percent's death in 1996. As the deputy chief of Golkar, Indonesia's ruling party, Tutut has set herself up as a potential king-maker when Suharto finally leaves office. Some feel she has a shot at becoming the next president, a prospect many Indonesians shudder to contemplate. People have tired of the Suharto clan's nepotism; should Tutut succeed her father, the practice will almost certainly continue.

"Suharto is brilliant, like a chess player," says one executive member of the Indonesian Mining Association. "[But] his children are a sore spot. They are rich enough, but Suharto will not listen to criticism of them. He always changes the subject." Suharto clearly demonstrated his blindness in an autobiography published in 1988. He claimed, unconvincingly, that his children were normal citizens, with no special privileges. "None of my children have been pampered. Not one. . . . On the contrary, they take a low profile and do not feel or act as if they are children of a President. Instead, they are actually self-effacing. . . . It is only right that they live as ordinary members of society, facing the realities of life." In a country where poverty is the norm, and people live in squalid conditions, drawing their drinking water from filthy canals a stone's throw from the palatial homes of the rich, Suharto's claims are both preposterous and insulting.

While no one dares criticize the president, the resentment towards his children and their self-serving business dealings is finally being vented in public. Their fortunes are potentially on the way down, depending on how Suharto's succession is played out. But Barrick did not contemplate this when it decided to reach out to Tutut. Although its advisors in Jakarta had given a number of interesting alternatives — including a scheme to join forces with the army and tap into its enormous pension fund — Barrick pursued only one course of action.

Tim Scott was cut out of the loop, having been summoned to Toronto and told to disgorge all he knew. But Munk still needed someone to make the introduction to Tutut. He sent two of his top guns — Neil MacLachlan, executive vice-president of investments, and Patrick Garver, Barrick's general counsel — to Jakarta to act as point men. They reached Tutut through one of her business associates, Airlangga Hartarto, the son of Indonesia's coordinating minister of trade and distribution. Tutut was informed of Barrick's desire to become involved in Busang. She was told that the Toronto company was willing to give one of her companies, PT Sumber Hema Raya, a contract to build some of the infrastructure at the site, assuming it ever went into development. Tutut agreed to help move Barrick's proposal through the political system when that became necessary. MacLachlan "handled the majority of the discussions, negotiations and ongoing relations" with Tutut's company, says a Barrick executive.

But moving the strategy into gear still required support within the Department of Mines itself. Since the rule-bending technocrats had already shown their support for Bre-X, Barrick sought to go a little higher up the food chain and engage the department's controversial minister, Ida Bagus Sudjana, a barrel-chested, high-caste Hindu from the Indonesian island of Bali. Airlangga, whose formal "association" with Barrick was kept a secret, was called upon to make that connection as well. His best friend and business partner, Dharma Yoga, was Sudjana's son.

A military general, Sudjana is a notoriously corrupt figure. He is known to play games with private industry, wringing last-minute concessions from domestic and foreign companies, squeezing them for cash, rationalizing that it is for the good of the country, when most people suspect that money is travelling into his own bank account. Sudjana surrounded himself with cronies and fixers, such as the unflappable Adnan Ganto, a cloak-and-dagger character who served as the minister's economic advisor, appointed by Sudjana's personal decree. Ganto wielded considerable clout. As an elite backroom deal-maker, he made friends in the Indonesian military by "speeding up" the purchase of Skyhawk fighter jets from Great Britain. Ganto was "the real authority" within the Department of Mines and Energy. Allegedly, his specific purpose was to direct contracts and projects to friends of the minister.

This wasn't unusual; this was the Indonesian way. All of Sudjana's predecessors had done it, albeit in a more elegant fashion, according to veteran civil servants. Ministerial interference was a time-honoured tradition within the department, something Sudjana made no attempt to hide. As he once explained to a legislative commission investigating charges of corruption within his department, he encouraged distributing "bits and pieces" of projects to personal associates. He defended the practice by insisting he gave only to "the younger generation, medium and small-scale entrepreneurs." He played the nationalist card, claiming that his frequent demands for unscheduled "royalty payments" and "administrative fees" from foreign resource companies were to "benefit the Indonesian people."

No one believed any of it. When he became minister of Mines and Energy in 1993, Sudjana tore up a number of major electricity and construction contracts and handed the work to companies in his own camp. Money flowed in and out of his own bank account, according to some reports. Reform-minded technocrats came to regard Sudjana with contempt. He always had his hand out, complained Kuntoro Mangkusubroto, who served under Sudjana as director general of mines. He knew nothing about mining. Eventually, members of the foreign mining establishment ridiculed the minister, as well. An executive with Inco described him as a mentally challenged "mafia type" with limited social skills.

This didn't stop Inco from cosying up to Sudjana during a junket he made to Canada in June, ostensibly to promote Indonesia's mining potential. Such trips are often arranged between Canadian mining companies and foreign administrators, perquisites that help maintain favour. Joining Sudjana was a group of advisors and technocrats, including a wary Kuntoro. Although Inco was the primary host, Barrick wheedled its way into the picture, taking the Indonesians on an aerial sightseeing tour over Niagara Falls, and throwing a lavish dinner party at a private Toronto club. Munk used that occasion to introduce Sudjana to one of his star directors, Brian Mulroney. The former prime minister greeted Sudjana warmly and expressed Barrick's interest in Busang. To drive home the point, Mulroney later sent Sudjana a letter, thanking him for coming to Canada, and again mentioning that Barrick, with its financial might and mining experience, ought to be involved with the development at Busang.

There was one last social event in Toronto. Before flying off to Europe for a series of meetings, Sudjana and his entourage attended a rather stiff reception and dinner at the Indonesian consulate on Jarvis Street. A number of other Canadian mining companies were invited to attend, and a spot at the head table had been reserved for Bre-X. David Walsh showed up and was surprised to find himself several places removed from Sudjana, while a representative from Barrick was seated right beside the minister. "I hardly had a chance to talk to Sudjana," he complained.

Walsh was way out of his element. It was all he could do to keep up with the political manoeuvring going on around him. He didn't know it at the time, but Barrick had already won Tutut's support. Now Peter Munk and his executive team were crafting some kind of mutually beneficial arrangement with Sudjana, which would give them the muscle to move in on Busang.

Barrick officials flatly deny that Sudjana was bought or that he was offered anything. They deny the existence of a Barrick–Sudjana juggernaut. But they can't adequately explain why the minister suddenly became so determined to push Bre-X into a one-sided deal with Barrick, and only Barrick. Sudjana dictated a dubious arrangement that would have slammed the door on all other suitors, dropping the lion's share of Busang into Munk's lap without benefit of an open auction process. To this day, Barrick maintains that Sudjana believed it was the only mining company capable of developing a large gold deposit in Kalimantan. Barrick spins a tale that it acted as a good corporate citizen. This is nonsense, meant to deflect attention from Barrick's own covert — and morally suspect — strategizing.

In July 1996, as Canada's ecstatic gold analysts were being squired around Busang, Barrick and its allies launched their squeeze play. The first formal order of business was putting the technocrats on notice that Barrick was the minister's choice to develop Busang. Kuntoro was hauled into Sudjana's office, located in a sombre, heavily guarded building in Central Jakarta, within sight of the sprawling presidential palace. "The minister asked me to support Airlangga," says Kuntoro. "It was their intention that Barrick become a partner in Busang."

Kuntoro was dismayed. Up to that point, the industry had been spared the kind of large-scale internal political graft that dogged other reaches of the economy. Indonesia's ruling class figured that mining was too capital-intensive, too long-term to bother with. Sudjana had initially shown no appetite for mining companies after becoming minister in 1993; he found it better to hijack other industries within his portfolio, those that offered quick cash flows. Sudjana had arranged deals for his friends in oil and gas, construction and hydroelectricity. But no one could resist Busang, not investors, not analysts, certainly not Indonesian politicians. Sudjana had decided to march into the booming mining scene with palms outstretched, waiting for a slap of cash.

The Contract of Work system, which most mining companies regarded as an effective, sober set of laws, was in danger of being subverted by the Barrick–Sudjana alliance. The idea that a politically connected company could make a secret deal and impose its terms on another company would not go down well within the industry. If Sudjana intended to lurch into the fray and begin demanding favours and special deals, the entire COW process, which the technocrats held so dear, would lose all legitimacy.

At first, Kuntoro refused to help the alliance. "I told them that it was none of my concern. I didn't want to get involved with their business. [But] the minister ordered me to introduce Airlangga to Bre-X." Two weeks later, Felderhof was summoned to Kuntoro's office and was surprised to find Airlangga Hartarto sitting on the director general's sofa. Kuntoro told Felderhof that Airlangga was

Barrick's agent, and that Barrick would like to discuss a joint venture. There were no numbers bandied about; it was just a get-to-know-you session. But Airlangga "came at it hostile," says Kuntoro. "He talked about Bre-X falsifying documents. He made threats. I think Bre-X understood what was going on. Barrick intended to become a partner in Busang."

Felderhof didn't get it; apparently, he failed to realize this wasn't a suggestion, but a command. "I was surprised [Barrick] came via Airlangga," he told me later. "I mean, why would they use Airlangga? [Barrick] had their own office in Jakarta." The reason wouldn't become clear until another meeting, convened by Airlangga in late July. This one was held over dinner. Francisco had flown in from Toronto to attend the meeting with Felderhof. Two heavyweights from Barrick, chief financial officer Randall Oliphant, and Pat Garver, sat on the other side of the table with their new "associate," Airlangga.

Once again, the Barrick side declared its interest in forming a joint venture to develop Busang. Airlangga continued to press hard, giving Felderhof the impression that he thought a deal should be finalized right then. Francisco countered that Bre-X wouldn't even hear an offer from Barrick until it had held discussions with other prospective partners, including Placer Dome and Teck.

Airlangga upped the ante. Towards the end of dinner, he mentioned that Barrick had enlisted Tutut's support in its quest for Busang. This caught Felderhof and Francisco completely off-guard. Finally, they began to realize that Barrick was mounting a "heavy-duty pressure" campaign to inveigle its way into Busang. "We found that Barrick was becoming, you know, came on quite strong," recalls Felderhof. "Like they wanted to do the deal and that they were the ones to do the deal with. The government was in favour of them doing a deal with us, okay?"

This was enough to make both Francisco and Felderhof sweat, although likely for different reasons. As Bre-X's new money man, hired to repair the company's neglected corporate affairs, Francisco had been completely occupied putting out administrative fires since joining the company in the spring. He had already worked the ten-for-one stock split. At the time, Bre-X shares were trading at a staggering $201, well out of reach of ordinary investors. Post-split, they were worth $20.10. Francisco also devised a shareholders' "rights plan" that made it impossible for Bre-X predators to mount a takeover in the open market. Finally, he negotiated a $7.1-million option payment to Montague Gold for the Central Zone and had started an investigation into concerns that an ugly ownership dispute loomed over the rest of Busang.

Now he was discovering first-hand that Bre-X had been completely outflanked on the political front and was in danger of falling into an unfavourable deal with a smarter, savvier rival. According to his former colleagues, it was at this point that Francisco began to regret ever having left his comfortable job back in Toronto.

Felderhof had far greater worries. He had thought that Bre-X could control its own destiny and milk the "world's biggest deposit" story as long as access to outsiders was restricted and the site remained closely supervised by de Guzman and his crew of Filipinos. Felderhof knew people would try sneaking in. But he hadn't counted on anyone breaking down the door. Barrick was pushing past the gates; there was nothing Bre-X could do to stop it.

Once Barrick had lined up its powerful political support, the time had come to put the information Kroll & Associates had dug up on Bre-X to good use. First, Barrick helped advance Yusuf Merukh's dispute with Bre-X. Rumours of a pending ownership dispute soon found their way to Canada and appeared in at least one analyst's report before documents were leaked to the press. Merukh had sent two formal letters to the Department of Mines, arguing he'd been shut out of the Southeast Zone. He claimed he should have been offered a minority ownership stake in the Southeast Zone, since he had a natural right to participate in claims found adjacent to the original Busang property. Merukh also insisted that Bre-X's interest in the Southeast Zone was the result of some ill-defined "confidential information" that, he said, belonged to PT WAM. Since he owned ten per cent of PT WAM, Merukh argued that the information that led Bre-X to the Southeast Zone should also benefit him. What's more, he alleged that he had an option to purchase another twenty per cent of PT WAM, which, if exercised, would give him thirty per cent of the Southeast Zone.

Merukh was a slippery character, widely dismissed by mining insiders as an irritant and an opportunist. The former leader of a small, Christian-based political party, who held a seat in Indonesia's limp House of Representatives for twenty years, Merukh was also ubiquitous in the country's mining industry. In the 1970s, he was practically the only Indonesian businessman to recognize the country's mining potential. Merukh seized control of dozens of concession areas throughout Indonesia; by the time the first big wave of foreign explorers showed up, he controlled thirty-four of Indonesia's top mineral prospects. He hadn't developed any of them; instead, he waited to sell them to foreign partners, while maintaining a minority interest in each site. This strategy had made him enormously wealthy and gave him an extraordinary degree of influence.

"The road to riches usually runs through Yusuf Merukh's office," moans one Canadian mining executive, who has had extensive dealings with the short, nervous Indonesian entrepreneur. "He was my partner in a project for five years. One of my biggest mistakes was doing business with him. He always wanted more money from me. He'd use every excuse. He'd say, 'Give me money, my daughter is sick.' I used to lie awake at night thinking of ways to assassinate him. He's the most odious joint-venture partner known to man."

The technocrats held a similar opinion of Merukh. Kuntoro described him as "very dangerous and manipulative. He was *persona non grata* in my department." Still, Kuntoro was disturbed by Merukh's letters outlining his claim on Busang. "I could not ignore them," he told me. "The case was very complicated, too complicated for me to comprehend. There was some evidence that Yusuf had a legitimate complaint."

Kuntoro was faced with a dilemma. He wanted to help Bre-X

secure a long-term COW for the Southeast Zone, but if Merukh's claims were real, the actual development of Busang could be delayed by years of legal wrangling. Handing a flawed contract to Suharto to sign, only to see it blow up in his face later on, was not something Kuntoro could ever contemplate doing. "I did not want to embarrass the president," Kuntoro said quietly. While the rules may have been bent here and there in order to accommodate Bre-X, Kuntoro, ever the Machiavellian, was determined that the system should at least *appear* to be working smoothly.

Meanwhile, the Barrick alliance was pressuring Kuntoro to boot Bre-X off Busang. The ownership dispute over Busang could be used as an excuse, the alliance suggested. So could Bre-X's other transgressions, including its hyper-aggressive exploration activities and the altered COW form it had filed with the SEC. "Barrick, via Airlangga, with the minister's support, wanted me to revoke Bre-X's COW application," Kuntoro admitted to me. "I said no. It had been approved in principle by parliament. I also pointed out that to maintain the trust of foreign investors, we must be seen to be following the rules."

Kuntoro told Sudjana the maximum that he could do was revoke Bre-X's SIPP, the preliminary investigation permit that allowed the company to conduct surface exploration. This, Kuntoro reasoned, "would say, 'Hey, Bre-X, here's a signal to consider Barrick as a partner.'" Although such a move would go against his principles, Kuntoro felt it was the "correct" thing to do since it was the farthest he could legally go to satisfy his minister.

On 15 August, Kuntoro cancelled Bre-X's SIPP, the result, he explained to Bre-X, of some vague "administrative problems." Felderhof told me later that this was "absolute bullshit." The SIPP "was cancelled for no reason at all," he said. "Everything, all the paperwork, was done properly."

This wasn't true; Bre-X knew it, Barrick knew it, and Kuntoro knew it. Losing a SIPP was serious; it was, in fact, the first time since the COW system was introduced that such a fate had befallen a mining company in Indonesia. And yet Bre-X did not share this vital piece of information with regulators, shareholders or anyone else, at the very moment that Bre-X was about to be listed on the NASDAQ, one of America's premier stock exchanges. Instead, the company pumped out a new resource calculation of 46.9 million ounces of gold for Busang. "This resource calculation has been independently verified by Kilborn Engineering," Bre-X claimed. This, of course, was misleading. Kilborn had not "verified" anything; it had only taken a series of numbers from Bre-X and come up with an estimate. The phrase was used again and again, and yet, curiously, Kilborn never insisted that Bre-X clarify it.

Not only was Bre-X hiding negative information from the public, its chief officers began dumping stock at a furious rate. According to insider trading reports, David Walsh unloaded 298,700 Bre-X shares between 8 August and 27 August, 1996 — about four per cent of his declared holdings — for a $7,465,735 windfall. Jeannette Walsh sold fifteen per cent of her shares in the same period, making $7,497,485, and dropped another 150,000 shares a few weeks later. Felderhof liquidated 14. 6 per cent of his Bre-X shares in August and continued to sell throughout September. John Thorpe, the company's part-time accountant, dumped thirty per cent of his Bre-X stock in August. Francisco, meanwhile, sold 50,000 shares on 1 August, just a few days after his nerve-racking meeting with Barrick in Jakarta. Francisco would liquidate his entire position before the end of the year. (Paul Kavanagh, who owned more than two million Bre-X shares, held onto his position, and later donated 11,500 shares to the Toronto Symphony Orchestra.)

While these trades were duly recorded with various securities regulators, as was required by law, the information would not be publicized for months. In the meantime, Bre-X continued to act as though nothing were wrong. In early September, the company announced that J.P. Morgan and Republic National Bank would be acting as its "financial advisors." This suggested that Bre-X was preparing to negotiate a joint-venture agreement after holding an auction; in fact, the company was preparing for an inevitable showdown with the only company really in the running for Busang,

Barrick Gold. A Barrick representative says that J.P. Morgan was "telling Barrick that Bre-X had no plans to hold an auction." The two sides met in Toronto in September and, according to Felderhof, discussed a fifty-fifty deal. No agreement was reached, because J.P. Morgan and Republic Bank refused to sign off on the proposal.

Hearing this, Airlangga went back to Kuntoro and complained that Bre-X was holding out, that it wanted to retain a larger share of Busang. This, he said, was impossible. The mining bureaucracy was instructed to get tough. When Sudjana's chief henchman, Adnan Ganto, appeared on the scene, the technocrats knew that something bad was brewing. Ganto approached Rozik Soetjipto, one of Kuntoro's close associates, and asked him if Bre-X's approval in principle to develop the Southeast Zone could be shredded. Rozik refused. "I said, 'You need a very strong reason to do this. Approval in principle is a strong legal instrument. In Bre-X's case, I don't see a reason.'"

The technocrats' fealty to Bre-X eventually cost them. On 17 October, Sudjana issued Ministerial Decree No. 1409, stripping Kuntoro's office of its responsibility to process all COW applications. "Outsiders are not entitled to know the reasons," Sudjana told reporters. Kuntoro and his allies were furious. They would not let the minister destroy their fiefdom without a fight. They had information that could harm Sudjana and Ganto; in time, they would use it, and with devastating effect. For the moment, however, they sat back and watched the dramatic events unfold.

In Canada, rumours of an impending ownership battle continued to circulate on the Street, but the information was vague, and the public remained ill-informed. Not wanting to spook investors and precipitate a sell-off, most analysts either played the stories down or kept quiet. Douglas Goold, an investment columnist with *The Globe and Mail*, had been informed in August that Bre-X had lost its exploration permit and was in danger of losing its grip on Busang, but he sat on the blockbuster information for more than a month before calling the Indonesian consulate in Toronto to confirm. In the meantime, the *Globe* continued to heap praise on Bre-X and Walsh. On 28 September, the newspaper's "Streetwise" columnist, Andrew Willis, repeated one of Egizio Bianchini's most notorious slips, that "good [mining] properties tend to be found by guys with checkered pasts." Willis fashioned this clanger into a ringing endorsement of Walsh, even though he had never met the man, nor had he spoken with him. The Bre-X founder "had some financial skeletons in his closet," Willis wrote, "but winners knew that he had the ability to identify promising properties." Willis did not elaborate. He couldn't, because his statement was patently false. Walsh had never identified a promising property in his life. "I just took what Egizio told me and went with it," Willis later admitted. "I didn't check up on Walsh at all."

It was clear that none of the other journalists covering Bre-X at the time had conducted any background checks on Walsh either — a disheartening situation, but not particularly surprising. Business journalists, especially those writing for daily newspapers, frequently depend on investment analysts for information. Analysts are presented as independent sources, with up-to-the-minute, impartial information. This is a dangerous misrepresentation. Many analysts are completely biased; they often hold shares in companies that they discuss with reporters, and their investment firms often underwrite the same stock. As the Bre-X saga plainly illustrates, an analyst's "research report" may simply consist of material handed down from the company he is following, rewritten to appear original. This rather salient fact is rarely passed along to newspaper readers.

While *The Globe and Mail*'s Bre-X coverage was for the most part sloppy and unsophisticated, the newspaper redeemed itself on 4 October by finally reporting the SIPP cancellation. The news sent Bre-X shares tumbling for the first time since its unorthodox assay methods were revealed the previous spring. Bre-X stock fell two dollars, to $24.55, prompting the TSE to halt trading for several hours until Bre-X issued a statement insisting that everything was in order.

Bianchini was caught off-guard. Apparently, the analyst billed as closest to the action had failed to see this coming. Under pressure from clients and colleagues to issue a response, Bianchini circulated a confidential memo. Instead of urging caution, he recommended investors accumulate Bre-X stock, dismissing allegations from the Indonesian consulate that the company "did not play by the Indonesian rules." This, he argued, was merely "a ploy... to put pressure on BXM." Although he noted that "10% of Busang may be at risk," Bianchini insisted that the deposit "will get much bigger than it currently is. The growth will far outstrip the potential cut in ownership." The following day, he wrote a positive update for public consumption, again recommending Bre-X stock as a top-rated buy, and predicting the Busang deposit contained more than sixty-one million ounces of gold, exceeding the company's own pronouncements.

Two weeks later, the *Globe* reported on Merukh's claim, and Bre-X shares fell another three dollars amidst heavy trading. Since the beginning of the month, Bre-X's market value had dropped a stunning $1.3 billion. While this sent loyal Bre-X investors into a tizzy, it must have thrilled Barrick. The farther Bre-X fell, the easier the takeover.

Munk couldn't contain himself. According to a source within Barrick, he doggedly lobbied the Canadian government to help push Bre-X to the table. When Canada's new ambassador to Indonesia, a career diplomat named Gary Smith, presented his credentials in Jakarta, he "hammered the Bre-X issue, arguing that Bre-X was a good company, that it had found [Busang] fairly." This got back to Munk, who happened to be at a dinner with Art Eggleton, then Canada's minister of trade. Eggleton was making a speech that night; apparently overcome by exhaustion, he fell off the dais and broke his leg. According to an amused Barrick source, Munk "pursued [Eggleton] to the hospital, chewing his ear off about Busang, saying the government should revisit the issue. That should tell you something about Munk's singlemindedness."

Using his best diplomatese, Ambassador Smith avoided commenting on the incident when I met with him at the Canadian embassy in Jakarta several months later. "I think here, I don't want to go over all the details of the past," he said. "When it comes to a commercial negotiation, we have to leave it to the companies to decide what's in their best interests." When I asked the ambassador if Munk had ever directly approached him about Busang, Smith paused and then ducked. "Who told you that? I don't, uh, I don't recall that that was an issue of contention."

Munk didn't stop there. He prevailed upon George Bush, his high-profile — and highly compensated — board member, to write a love letter to President Suharto, extolling Barrick's merits. Bush was happy to oblige. The letter, written on Bush's private stationery and dated 20 September, 1996, quickly gets to the point:

Dear President Suharto,

I want to thank you for your thoughtful letter. I hope my travels will soon bring me back to Jakarta so that we might meet again.

Earlier this week, I was in Elko, Nevada for a meeting of the Barrick Gold Corporation's International Advisory Board (to which I am a Senior Advisor) and had the opportunity of touring the company's vast mining operations there.

Because Barrick Chairman Peter Munk advised me of their interests in a major gold development in Kalimantan, Indonesia, I simply want to take the liberty of telling you how impressed I am with Barrick, its visionary leadership, technological achievements, and great financial strength.

I can recommend Mr. Munk and Barrick with no reservations whatsoever.

My respects to you sir and my warmest best wishes.

Sincerely,

George Bush

While Barrick executives now acknowledge this letter was sent to Suharto, they claim "it is the only Bush involvement...in the matter." Even so, Barrick had demonstrated its ability to penetrate deep inside Jakarta's corridors of power, something Bre-X had never even bothered to attempt. Some observers argued that Walsh and Co. were in over their heads, that they were naive and stupid for trying to go it alone. Yet when Bre-X did go and buy some protection of its own, the company was criticized for doing business with corrupt forces and cutting a "cynical" deal.

Bre-X was simply following Barrick's lead when it approached one of the Suharto kids. The difference was that Barrick had kept its deal with Tutut quiet. Bre-X took the opposite tack, proudly announcing it had formed a "strategic alliance" with PT Panutan Duta, part of a private conglomerate controlled by Suharto's eldest son, Sigit Harjojudanto. Panutan would "assist in administrative, technical, and other support matters within the Republic of Indonesia, including the identification of issues concerning the acquisition, exploration, development and production from mineral resources and other interests." In exchange, Panutan would receive a ten per cent interest in Busang, along with US$40 million, paid out over forty months.

Although the deal would dilute their interest in Busang, Bre-X shareholders responded favourably to the arrangement. Most Bay Street analysts suggested it would help Bre-X overcome its problems in Indonesia, and Bre-X shares moved back up to $24. In Jakarta, people just laughed. While Tutut was a considerable force in politics and business, Sigit had little credibility. Content to ride along on his younger brothers' coat tails, Sigit seldom won any accolades from his father, who doted on his siblings. His one individual effort at entrepreneurialism — the creation of a national sports lottery — was sideswiped by intense criticism from Muslim groups and was dissolved in 1993.

Politically impotent, despised by prominent Indonesian nationalists, ignored by his own father, Sigit did nothing to help Bre-X's case. "Sigit is a complete idiot," says Barrick's Tim Scott. "He has a record as a dissolute and a gambler. He was the wrong guy to pick." Indeed, his involvement only tilted the balance in Barrick's favour. With the meddlesome technocrats emasculated, at least temporarily, and Tutut sitting in his corner, Sudjana felt free to dictate how the development of Busang would proceed.

The ultimatum came on 14 November. Walsh and Munk were

asked to present themselves at Sudjana's ministerial chambers. Walsh showed up with Felderhof, Francisco and representatives from J.P. Morgan and Republic Bank. Munk was the only person from Barrick to attend, which suggested that he already knew what Sudjana planned to say and that there would be no negotiation.

The men gathered around a conference table down the hall from Sudjana's office. Kuntoro and Rozik were there, along with Ganto and Umar Said, Sudjana's bellicose secretary general. Sudjana strode in majestically, sat down at the head of the table and began reading out loud, in halting English, from a prepared statement. Barrick, he said, would have seventy-five per cent of Busang. Bre-X would have twenty-five per cent. The Government of Indonesia would appreciate a ten per cent interest in the deposit. It was up to Barrick and Bre-X to decide how to arrange for that. And that was it. Sudjana got up and left the room. After a few moments of silence, Said spoke up, telling the two parties they had eight days to come to a complete financial agreement or the government would consider expropriating Busang.

This was insane; the fact that the terms were hopelessly one-sided was almost beside the point. Barrick had a fiduciary responsibility to its shareholders to conduct due diligence on the property before concluding any kind of deal. The company hadn't set foot on Busang since Paul Kavanagh's visit, and that was eons ago. Then there was the Merukh dispute; it still hadn't been resolved. The minister's directive obviously indicated he couldn't care less about Merukh's claims. That issue had been a smokescreen after all. Finally, Sudjana had given no indication about what should be done with Bre-X's other Indonesian partner, Haji Syakerani.

Back in Canada, investors reacted with either horror or delight. Some saw right through the positive rhetoric spun by Bre-X and Barrick. Both companies tried to mollify their shareholders by assuring them that a mutually beneficial agreement could be reached quickly. Bre-X was particularly hard-pressed to sound cheerful, since it had lost all but a slim slice of Busang. In an effort to buoy its stock, which threatened to fall through the floor, the company released yet another resource estimate endorsed by Kilborn, alleging that the property contained fifty-seven million ounces of gold. Even better, argued Barrick's supporters. Munk's company appeared poised to grab control of a bottomless treasure chest.

Rather than question his methods, or investigate how Munk might have managed to advance his position inside Indonesia so quickly and decisively, most of Canada's media stood back and applauded. The most egregious example of mindless cheerleading came from the weekly *Maclean's* magazine, which promptly pasted Munk's image on its cover, proclaiming him as the "King of Gold." Inside, business writer Jennifer Wells, in an uncharacteristic fit of sycophancy, wrote that Munk "has played his imperial hand beautifully." Compared to David Walsh's bumbling, "the moves of Peter Munk have been absolutely stealth-like. . . . Through it all, Munk has remained his sleek, reserved self." *Maclean's* columnist Peter C. Newman, one of the most prolific chroniclers of Canada's business elite, also got in on the act. He panted over Munk, calling him a "vaguely supernatural" leader, endowed with the "sophistication of a skilled swordsman, knowing precisely when to feign and when to thrust."

It was hard to decide which was more sickening: Barrick's tainted victory or the media's celebration of it. The real story, it seemed, would never emerge inside Canada, where people are willing to accept the most foolish tales. The truth would reveal itself on the other side of the planet, or so I hoped.

A week before Christmas, I arrived in Indonesia, where fires were burning.

11

THE YEARS
OF LIVING
DANGEROUSLY

*Who wants to live like us? No one — that is, if they
have any choice. But everyone knows, too, that we're
one part of something immoral. To talk of changing the
situation, that's nonsense. The only thing that's going to
change the situation is a miracle.*
— From the banned Indonesian play *Time Bomb,*
by Riantiarno

WASN'T PREPARED FOR IT, INDONESIA. I was
entering the country illegally, without the required
visa identifying myself as a working journalist. I'd
been warned not to attempt this. Foreign reporters caught posing as
tourists were routinely kicked out of the country.

A last-ditch effort to obtain the proper documentation at the
consulate in Toronto had been useless. After informing me that her
country was "not corrupt," the consul insisted that I promise to not

write a word about the politics of the place. Only then would she see if she could rush my visa application through the system. I told her I was simply after the truth. It was important that I arrive in Indonesia before Christmas; there were other Canadian journalists trying to get in the country, including Douglas Goold from *The Globe and Mail*. He had applied for a visa around the same time as I had, but rather than sneak into the country he apparently decided to play by the rules. He waited two months for his application to clear.

I was determined to be the first on the scene to report on the Bre-X affair, and so I landed in Jakarta on 19 December, 1996, minus the appropriate stamp in my passport. Just before landing, a flight attendant handed me my customs card. I checked "Tourist" under the question "Purpose of Visit." I had dressed the part as a precaution, wearing chinos, T-shirt, ballcap and sneakers, rolling my one business outfit into a canvas knapsack. Another inconspicuous piece of luggage contained the various tools of my trade: tape recorder, empty cassette tapes, notepads, a few "seditious" books about Indonesia that had been banned by the local authorities. I prayed that I wouldn't be asked to open it. "If they find these books on you, you'll be on the next flight to Singapore," Kurt MacLeod, the promoter with the white beard, had told me, just before I left Toronto.

The first order of business was getting past a gauntlet of stone-faced customs officials and heavily armed soldiers who were positioned at various points throughout Jakarta's Sukarno-Hatta Airport. I tried to look a little bored and world-weary, but my stomach was tied up in knots. "What do you plan to do here?" the customs man asked in English, glancing at my luggage. "Relax," I said. He looked at me, eyes narrowing. "I mean, I intend to relax." I smiled at him, and he waved me along.

As I later realized, my first cab ride from Sukarno-Hatta to downtown Jakarta neatly encapsulated the essential Indonesian experience, in just forty illuminating minutes. The trip began pleasantly enough, along a modern four-lane highway, smooth as silk and lined with neatly trimmed shrubbery and lovely, beckoning palm trees. The air was soft, warm and moist, a wonderful contrast to the stale, frigid atmosphere inside the Boeing 747 that had carried me here from Amsterdam. We passed a large field crowded with hundreds of brand-new automobiles. These, my cab driver said with a sneer, were Tommy Suharto's Timurs, the clunky, Korean-built "national" cars that no sane consumer touched. A few minutes of silence passed before the cab rolled up to a toll booth. The driver asked for money. "For Tutut," he said, referring to President Suharto's eldest daughter. One of her companies owns this private roadway.

We stopped at two more toll booths, and I handed over more soiled Indonesian bills. The scenery crumbled the farther into the city we travelled. Parts of the highway were swamped by heavy rains; children stood atop muddy dikes, fishing for carp with bamboo poles. The road narrowed; suddenly, my driver was leaning on his horn. The lanes filled with ramshackle buses farting black clouds of exhaust. Motorcycles roared all around us. The trees and bushes had disappeared; dilapidated, tin-roofed huts took their place. Mercifully, our car escaped by speeding up a ramp and throttling out onto an open causeway.

We were fast approaching the city centre. Construction cranes dotted the horizon, an impressive sight that reminded me of the building boom that had transformed western Canada's largest cities two decades earlier. Beautiful new skyscrapers line the route. We passed by the US$300-a-night Shangri-La Hotel, where the Bre-X team was camped out at that very moment. The driver pointed excitedly to a glittery new franchise restaurant next door. The Fashion Café, freshly imported from the West, now caters to Jakarta's young affluent class. Down the street was a brand-new Planet Hollywood. I checked out the prices a couple of weeks later. I couldn't afford to eat there.

The cab swerved onto a darkened side street. We passed over a stinking canal. Stray dogs were pawing through fetid piles of refuse on the sidewalks. Men and women crouched next to crooked little food carts, slurping steaming noodles from plastic bowls. We passed a movie theatre, where a huge, hand-painted image of a pistol-packing Bruce Willis glared down from a giant billboard. Finally, the driver pulled up to a sad-looking budget hotel — favoured, it turned out, by boozy Slavic businessmen. The driver demanded 40,000 rupiah, about $23, although the dashboard meter indicated I owed him much less. He insisted, refusing to open his trunk until I sheepishly forked over the case. He slammed his door shut, gave me a dirty look and sped off.

Disoriented, I stumbled into the darkness, nervously patting my pockets, worried that I might have left something behind in the back seat. A young Indonesian man lurking beside the hotel approached and asked me if I wanted a girl for the night. "Very young," he said. "Very clean. You want dope? Hey mister, whatever you want, I can find it for you."

I spent the next few days trying to shake jet lag and acclimatize to my strange new environment. Reading local newspapers helped; there are three English-language dailies in Jakarta, and while they all pulled their punches when it came to discussing the president, the rest of their coverage seemed relatively thorough, if a little depressing.

It had not been a good year for Indonesia, although Suharto and his acolytes claimed the nation continued to make great progress. A series of Stalinesque "five-year development plans," introduced in 1969, had culminated with a number of "grand achievements." But what did that mean? The country no longer imported rice. The average income had jumped from pennies per day to a couple of dollars. Infant mortality had been halved, and life expectancy had increased to sixty-three years. Cold consolation to the estimated thirty million Indonesians still stuck in poverty, forced to drink water from sewers. Meanwhile, the Suharto kids had become filthy rich, accumulating assets worth an estimated $4 billion.

Suharto had made a few token gestures to narrow the gap between rich and poor. Announcing the country's 1997 budget, he promised Indonesia's six million civil servants a ten to twenty per cent pay hike. Entry-level bureaucrats would see their monthly salaries rise from US$90 to US$110; their most senior colleagues would enjoy a US$28 boost.

A few days before Christmas, Suharto decreed that "wealthy individuals and companies" earning more than US$60,000 annually would be forced to contribute two per cent of their net profits to a special poverty fund. Eligible taxpayers were informed they "must transfer their donations directly" to a bank account held by one of Suharto's many "charitable" foundations, the Yayasan Dana Sejahtera Mandiri. The *yayasan*'s treasurer happens to be Bambang Trihatmodjo, the president's very wealthy second son.

Suharto's decree was greeted with cautious scepticism. Most observers said they supported a fund for the poor, but worried that it could be abused. "I'm afraid [the fund] will not be useful if the receivers do not have specific plans in mind and throw the money around for nothing," noted an executive from a major cosmetics company. Political analyst Hartojo Wignjowiyoto went further, questioning the integrity of the *yayasans*. "Usually, money is not received or allocated properly. The people who enjoy the money are actually not the target groups but the people or institutions that run the allocation mechanism."

It was also pointed out that the president had no legal right to demand that citizens make payments to his foundation. Suharto shot back, warning that those who failed to submit money to the new *yayasan* would be "morally chastised." He called it a "duty, based on the state ideology *Pancasila*. In this country, those who are able should be willing to help the weak. Just mark those business people who refuse to donate two per cent of their profit. There's no need to lash out or be upset. Just mark their houses with flags or whatever. They should be ashamed then."

Dominating the media's annual reviews were summaries of dozens of civil disturbances, the result of blatant military political finagling, judicial corruption, ethnic and religious strife. Most incidents had ended in terrible violence. Suharto had started the year with a national address asking for calm, saying that "each day, each month that we pass without a major upheaval is a national achievement." Sadly, his words had gone unheeded.

In March, the *Jakarta Post* reported, one person was killed, "dozens" of people were injured, and "scores of houses and cars damaged" when soldiers rampaged through a small town in North Sumatra. The attack was believed to have been led by members of a cavalry unit, seeking revenge for the death of a colleague who had been stabbed by a local hoodlum. Later that month, a gang of disgruntled youths ran amok through the capital of Irian Jaya, following reports that the leader of a local separatist movement had died in a Jakarta prison. Four people were killed in that incident; a marketplace run by non-Irianese traders was torched.

A muckraking journalist who uncovered a large case of political graft was beaten to death in his home by an unidentified group of thugs. Another journalist was found floating in a river, his skull bashed in, after a story he wrote angered East Timorese rebels. Two thousand people rioted in West Kalimantan after a Dayak man was allegedly abducted by the militia. Ten thousand students rampaged in South Sulawesi when local administrators tried to raise bus fare by ten cents. At least forty convicts escaped from prisons in Irian Jaya, and one hundred buildings were destroyed, after a mob of unemployed men took to the streets to protest their situation.

The list went on and on. But the worst rioting by far took place right in Jakarta. In June, eight thousand supporters of Megawati Soekarnoputri, daughter of former president Sukarno, clashed with police outside a major railway station, situated in the heart of the city, just a few blocks from my seedy hotel. The group was protesting the government's ham-handed putsch of Megawati from the leadership of the Indonesian Democratic Party (PDI), one of only three "legal" political parties left in the country. Megawati and the PDI, a coalition of moderate Christian, Muslim and ethnic organizations, enjoyed wide support from labour unions and younger Indonesians, including several million who would soon be of voting age. The growing movement had represented a serious threat to

Suharto, who had ordered that Megawati be deposed. This stirred bitter memories of Suharto's infamous rise to power, when he branded Sukarno a Communist sympathizer, booted him from office and launched the country into a decade of bloody purges.

"My father's spirit is still alive in the people of Indonesia," Megawati said defiantly. Suharto obviously agreed and took extreme measures to suppress it. Dissenters were punished. Students caught trying to form alternative political parties were jailed. Activists who refused to subscribe to *Pancasila*, the state ideology, were tried for subversion. One unfortunate creature was imprisoned for "insulting" the president; he'd suggested Suharto was no longer fit for office.

Megawati was more careful; she never made "illegal" statements and therefore could not be accused of subversion. It hardly mattered. In July, operatives from Suharto's ruling Golkar party installed one of their own as PDI leader. This sparked more massive demonstrations in Jakarta; thousands of protesters attempting to storm PDI headquarters were confronted by militia men and private security troops brandishing rattan sticks, iron bars and cannisters of tear gas. At least five people died in the rioting, and 150 were injured. Another twenty-three people were reported missing. Some of the rioting was captured on tape and broadcast to appalled audiences around the world.

Following a public investigation into the incident, Indonesia's National Commission on Human Rights issued a report that obliquely criticized the government's role. The commission, sanctioned by Suharto and partly funded by the Canadian government, concluded that serious human rights violations had taken place, including violations of the rights of assembly, freedom from fear and the right to life. But if Indonesians thought that this token bit of public relations was going to put an end to the violence, they were sadly mistaken.

The battle lines merely shifted; Muslim extremists began hunting Christians. Upset over a "mere" five-year jail sentence given to a man accused of blasphemy and slander against a respected Muslim cleric, a mob of angry devotees assembled outside a courthouse in East Java, waiting for the man to emerge. When rumours spread that the convicted man was actually hiding in a local temple, the mob went on a rampage, setting fire to twenty churches. Five Christian worshippers, including a disabled woman, burned to death. The courthouse, an orphanage and an unspecified number of schools and cinemas were also destroyed. It was the worst anti-Christian violence in years, but it didn't end there, as I discovered.

More churches were torched over Christmas, this time in Tasikmalaya, a city 150 kilometres southeast of Jakarta. Rumours that three Islamic teachers had been savagely beaten by police officers sent militant Muslims into a frenzy. A crowd gathered outside the police station and began throwing rocks. The *Jakarta Post* reported that "a group of unidentified people [then] told the crowd to attack churches and properties belonging to people of Chinese descent." After destroying thirteen churches, the mob — led by students from an Islamic boarding school — looted and then fire-bombed eighty-nine shops, twelve police stations, six banks, four factories, four schools, three hotels and one Hindu temple. About five thousand terrified Chinese were forced to seek shelter inside government offices until the rioting finally subsided.

Images of the destruction were broadcast on television the next day. The city looked like a war zone; overturned cars were lying about the streets, burned out and still smoking. Everything was charred. Broken glass littered the sidewalks. Then Suharto took over the airwaves, preaching calm. "Religious communities need to exercise self-restraint," he said. I wondered if anyone was listening. Rather than attend a local church service, I stayed inside my hotel room and watched the Pope deliver mass from St. Peter's Square.

Rozik Soetjipto looked at me anxiously before sitting down behind his desk. "There are those who think I say too much," he said. "But I think it is better people know just what is going on here." I couldn't believe my luck. No one outside the corridors of power knew how Barrick had muscled its way into Busang. And now one of the chief technocrats inside the Department of Mines was about to spill the beans.

I spent over an hour with Rozik. He was a most cooperative source — patient, polite, never insisting we go off the record. I was somewhat surprised when he agreed to let me snap half a dozen photographs of him at his computer. More important, Rozik helped set up a meeting with me and his boss, Kuntoro, who had refused all interview requests since Sudjana stripped him of his power. These interviews provided part of the foundation for a long article I later wrote for *Canadian Business* magazine, a piece that raised eyebrows among mining industry veterans worldwide, enraged Sudjana and his cronies, and deeply upset Munk and his team of executives at Barrick Gold. (One Barrick official — a lawyer, in fact — informed me the story "was wrong and it was libelous." Barrick, however, did not pursue this claim in any formal manner. The same Barrick official wrote a confidential, 8,500-word assessment of my *Canadian Business* article, attacking the motives ascribed to the company in the article, while also confirming a number of important details, and withholding any comment on other sensitive revelations. This reaffirmed my faith that the technocrats spoke the truth about Barrick's campaign to grab control of Busang. All Barrick-related details in this book have, in fact, been corroborated by Barrick employees.)

Essentially, Rozik and Kuntoro explained how Barrick landed its sweetheart deal for Busang. They described how Airlangga and Ganto had acted as their Indonesian agents, how Mulroney and Bush had sent letters on Barrick's behalf, and how Suharto had been misled by a Sudjana–Barrick alliance. They both characterized Sudjana as a crook, more interested in helping himself than his country.

Of course, they were hardly objective. Rozik and Kuntoro were decidedly in Bre-X's camp. Although neither man had much to say about David Walsh ("an unknown") or John Felderhof ("a bushman"), they were grateful that Bre-X had stimulated so much interest and activity in the local mining scene. And they were furious with Barrick and Sudjana, alleging they had conspired to subvert a well-regarded administrative system. "You can imagine how we feel," Rozik told me. "The minister had never given any clear explanation for forcing this [agreement between Bre-X and Barrick]. We were proud of our system. How can we face the Canadian mining industry now? Yes, Bre-X made mistakes. They were greedy and tried to inflate the value of their shares before they had received a Contract of Work. Perhaps they should give a percentage of the deposit to the local community. But there are proper ways to arrange for this."

Rozik and Kuntoro were worried that Sudjana's interference might set an ugly precedent, which would divert new investment away from Indonesia. The battle for Busang was now a major news item across the archipelago. Everyone was waiting with bated breath as Bre-X and Barrick locked horns and tried to come to terms with Sudjana's startling declaration. The first deadline had passed. So had the second. It appeared that Bre-X was dragging its heels, so Umar Said, Sudjana's assistant, began applying more pressure, telling reporters in Jakarta that the Calgary company might have misrepresented itself to the SEC in Washington. "The government is checking to see . . . if this is criminal," he intoned.

More stories came out, some based on tidbits picked from Barrick's Kroll report. Bre-X had made false claims to the Indonesian government regarding its ownership of Busang. David Walsh had made a false declaration on a visa permit, referring to himself as an Indonesian citizen. Another application stated he was a "geologist." Meanwhile, Said continued to thrash about, threatening to revoke Bre-X's claim on Busang if it didn't reach a final agreement with Barrick immediately. Finally, a week before Christmas, the two companies submitted a secret proposal to the government. Now the ball was in Indonesia's court. And this unleashed a whole new set of problems.

Few Indonesians had even heard of Busang prior to December 1996. But now that the fabled property was in the news, people began to wonder why two Canadian companies were fighting over it. Prominent nationalists argued that Busang's awesome riches belonged to Indonesians; it was they, not Bre-X, not Barrick, who should have it.

Amien Rais, a noted political analyst and chairman of the moderate Islamic organization Muhammadiyah, told reporters that Indonesia deserved more than ten per cent of the deposit. "Frankly speaking, when I read newspaper stories that they wanted to give two Canadian firms a ninety per cent controlling interest in Busang, my sense of justice was revolted.... If we do not have the funds and technology to tap the natural resources, why don't we leave those resources underground until we are capable of developing them? What's the point of us pressing the Canadian firms to solve their conflict quickly so that they can start working immediately to deplete our resources? We should keep them for our grandchildren in the twenty-first century."

His opinion was shared by Miendrowo Prawirodjoemeno, a professor of economics, who declared that the Indonesian government ought to be the majority stakeholder in Busang. Other nationalist-minded academics pushed the argument even further, stating that foreign companies threatened Indonesia's fragile unity, particularly in areas being considered for development by off-shore mining interests. Referring to rising tensions in Kalimantan, sociologist Loekman Soetrisno asked, "Can you imagine how much smaller the ethnic groups' living space would become once multinational corporations enter the region?"

None of these quite legitimate concerns had been raised in North America, something that bothered Arrif Arryman, another Indonesian economist who heads a Jakarta-based think-tank. "I don't wish to appear too critical of foreign companies and their shareholders," he told me, "but our national resources should be controlled by the state. That's in our constitution. Now it appears there is some kind of hanky-panky around Busang. We are entering a period of political uncertainty. This issue isn't helping matters."

Arryman pointed out that Busang represented all that was wrong with the present situation: A few highly placed political figures stood to benefit from a deal between foreigners, while the average citizen would likely see little or no benefit. It had always been so under Suharto. How else could a nation with such enormous mineral wealth allow such a glaring disparity between rich and poor?

Although Suharto refused to comment publicly on the Busang affair, there were indications that his daughter, Tutut, regretted that she'd ever become involved. Approached by reporters at a glitzy function in Jakarta, Tutut was asked to explain her relationship with Barrick. "What is this creature Busang?" she snapped. "I wish I'd never heard of it."

Tutut was deeply upset with Barrick. She had been assured that the company's final negotiations with Bre-X would be concluded quietly and without any fuss. Now she was being dragged into a national controversy. To make things worse, in December, a Barrick spokesman embarrassed her with some ill-conceived comments published in the *Far Eastern Economic Review*. David Wynne-Morgan, Barrick's director of corporate communications, admitted that Munk had signed a deal with Citra Lamtoro Gung, Tutut's private conglomerate, giving it exclusive rights to build roads and infrastrucure at Busang in the event Barrick gained control of the site. "The fact that it was owned by Tutut, we saw as an added advantage," Wynne-Morgan told reporter John McBeth. "Mr. Munk always likes to cover every angle." Wynne-Morgan added that Tutut would not get a free ride. Her company "would have to pay."

Neil MacLachlan, Munk's main man in Jakarta, was furious. "He was apoplectic with rage," says a Barrick employee. "You don't go around bragging that the president's daughter is on your side of a negotiation. And you don't tell people that Tutut has to cough up money for the privilege." Wynne-Morgan was fired a week later.

More revelations began to appear in the Indonesian press, severely damaging Barrick's insider campaign. Two widely read weekly Indonesian magazines — *Gatra* and *Forum* — raised allegations that Sudjana, Barrick's other key ally, had transferred US$23 million from a state-owned coal company and deposited the money into his personal bank account. Sudjana denied it, sort of. "The person who received and kept that money is the secretary general," he said, referring to his oily sidekick, Umar Said. "I never saw what that money looked like. Not one cent of it has ever been used." But that didn't explain why the money had been misdirected in the first place. The government later listed a number of conflicting reasons. The minister planned to use the cash to "help the poor." He was going to use it "to develop coal briquettes." The money was being held to avoid "wasteful spending."

These lame excuses impressed no one. President Suharto was reluctantly forced to make a public statement, in which he advised that the finagled funds "shall be returned to their rightful owner." Behind the scenes, Suharto was furious with his minister. "The president allowed Sudjana to stumble in public, without offering any support or advice," noted one technocrat. "Never before has a military man been treated like this."

It was open season on the minister. Sudjana's other dealings, including his involvement in Busang, became objects of speculation. Sudjana was called before an Indonesian House of Representatives committee to respond to a number of questions, including allegations that he and his cronies had conspired with a foreign company to pressure Bre-X into a lopsided deal. Sudjana referred the committee to Ganto. This created an uproar. Who, the committee wanted to know, was Adnan Ganto? What authority did he have to speak on behalf of the minister? "This ministerial advisor is well-known to the president," Sudjana insisted. "So he is not a vagabond. He has done a great service to this country." This failed to satisfy the committee, whose members insisted that Sudjana speak for himself. The embattled minister explained that his "chief mission was to spread projects fairly and evenly," among "the younger generation, among medium and small-scale entrepreneurs."

Emboldened, Indonesian media stalked Sudjana and Ganto day and night. One evening, the minister showed up at Ganto's colonial-style home in Menteng. Ganto and his wife, Tina, were celebrating their twenty-second wedding anniversary, and a number of high-profile politicians and businessmen were invited to the party. Outside, a team of private security officers kept reporters at bay. Wahid Rahamanto, a reporter with *Forum*, managed to get Ganto on his cellular telephone and peppered him with questions about Busang. Was it true that he'd been paid by Sigit and Tutut, for inviting them into the Busang deal? Was Sudjana's son, the developer Yoga Dharma, poised to win contracts at the site?

Ganto refused to answer the questions. "I know what I do," he said cryptically, and then threatened to call the police. Several days later, mining department flunkies advised one of Wahid's *Forum* colleagues that "there are large sums of money available to you, if you cooperate." Sitting down over coffee at my hotel, I asked the *Forum* reporter what that meant. "It was a bribe, of course," he told me. "If we stopped writing about the scandals, we would receive money. That does happen here."

The technocrats revelled in Sudjana's come-uppance. This didn't surprise me; I'd already figured out why Rozik and Kuntoro had agreed to speak with me. They were on a campaign to destroy Sudjana. "He will be gone in 1998," one of them told me, referring to Suharto's upcoming bid for "re-election."

His electoral success was a given. Rather than seek a mandate from the general population, Suharto merely had to win the approval of Indonesia's acquiescent general assembly, which consists of the nation's five hundred parliamentarians and another five hundred hand-picked appointees. It is Suharto's custom to shuffle his cabinet at the start of each five-year term, and everyone knew that Sudjana had hurt his chances of being re-appointed as minister of mines. "It would be best if Sudjana was removed now," one of the technocrats told me, "but Suharto is basically a king, and a king cannot admit to having made a mistake in appointing an incompetent to cabinet. We will just have to wait."

News of Sudjana's indiscretions had weakened Barrick's grip on Busang. Christmas had passed, and there was still no final agreement

between the two Canadian companies. This gave de Guzman a brief
reprieve. Terrified at the prospect of outsiders landing at Busang and
conducting a full due-diligence campaign, he had vowed to "pack up"
and leave the property if Barrick was given the green light, adding
that he would take the rest of the Filipinos with him. De Guzman
pleaded with Felderhof to continue resisting Munk and his allies. In
a memo dispatched to Felderhof in late December, de Guzman wrote:
"I know you can, as always, get the upper hand. If we let Barrick
physically in by drilling now, it is already the 'sign' of surrender."

The impasse had not gone unnoticed back in North America. A
group of disgruntled Bre-X shareholders were talking with Barrick's
rivals, including Placer Dome and Teck. Their goal was to stall
Barrick's one-sided deal with Bre-X and kick-start an auction for
Busang on the open market. The group's leader was Greg Chorny, a
retired lawyer from Aurora, Ontario. Chorny, who had recently
made a fortune in the gold business in northern Ontario, owned 1.4
million Bre-X shares, worth approximately $30 million. Although
he'd already sold 700,000 shares at an enormous profit, he was outraged over Barrick's attempts to seize control of Busang. He wasn't
mollified by Munk's promise that Bre-X would get fair value for the
property. "I don't care what Barrick offers," Chorny told me over the
telephone in January 1997. "I've lost all confidence in the company.
The deal they negotiated is open to all kinds of litigation."

Chorny's group had already engaged the Texas law firm Baker
and Botts to investigate whether Barrick had broken U.S. securities
laws in its attempt to seize control of Busang. Tom Ajamie, the savvy
Houston attorney whom I encountered months later at the "requiem
for Busang," was in Jakarta, chasing the same leads as I, albeit from a
much more comfortable hotel. Placer Dome had agents in Indonesia as well. The company's CEO, John Willson, made a clandestine
trip to the capital, meeting with a number of mining officials and
business leaders. His mission was to convince authorities that the
Barrick deal was not in Indonesia's best interests.

In early January 1997, Placer Dome made an attractive, politically sensitive offer to Bre-X, proposing a $6.2-billion merger and

reserving up to forty per cent of Busang for Indonesian interests. Chorny was ecstatic over the bid. "This is more like it," he said. "We've been trying for weeks to get other potential companies to come forward. Let the auction begin."

Placer Dome's offer took Barrick by surprise. Munk's company issued a press release, stating rather ingenuously that the "Indonesian Government supports" the Bre-X–Barrick joint venture. "The Government stated that Bre-X and Barrick should form a company and submit new Contract of Work applications. The Government confirmed an ownership split for the new company of Barrick and its partners 67.5%, Bre-X and its partners 22.5%, and 10% for the Government of Indonesia." This was misleading; the "Government" Barrick referred to was Sudjana. Thanks to his bumbling, he'd lost control of the Busang portfolio to his main political benefactor, Benny Murdani. The battle for Busang had taken on a completely new life. Forces far more powerful than Peter Munk and Ida Bagus Sudjana were gathering to decide Busang's fate.

The great irony, of course, is that Busang was barren. Although this wouldn't become clear for several more months, Barrick's initial due diligence — while limited — had indicated that there was very little gold at the site.

Barrick has consistently refused to disclose results of its technical audit of Busang core and bulk samples. The company argues that it is still bound by a "confidentiality agreement" it signed with Bre-X on 24 November, 1996. Given that Barrick was misled about Busang, it might seem odd to some people that it continues to respect this agreement. One explanation is that Barrick had reason to believe that Bre-X was misrepresenting Busang. In December, Barrick was given 135 samples of skeleton core from the property. The company flew the samples to a laboratory in Lakefield, Ontario, where they were assayed for gold. Only five of the samples indicated more than one gram of gold per tonne. According to Strathcona Mineral Services Ltd., the Toronto-based consultant eventually hired to determine whether Busang was legitimate, "the Barrick ratio of success . . . was higher than we achieved, but still far less than would be expected for the Busang deposit."

I later learned that Barrick air-freighted at least five tonnes of contaminated bulk samples from Busang to Lakefield, where they were tested for gold. While the results remain a secret, a source at Barrick told me that the company was concerned about the nature of the gold. "The official story is that Barrick did not proceed far enough with its due diligence," says the source. In fact, he says, Barrick knew that the gold contained in the samples was not analogous to its source. "Barrick knew there were rounded particles in there. [We] were concerned about free-milling [loose bits of gold]. Gold fell out [of the samples]." Rather than blow the whistle, Barrick kept these troubling findings quiet and continued with its takeover attempt. According to sources inside Barrick, Peter Munk was obsessed with seizing Busang. Getting control of the property was his first priority; determining its true mineral content could come later.

The company mounted one last public relations campaign, designed to reverse its negative perception in Indonesia and discredit its rival, Placer Dome. The company hired Brian Mulroney's former press secretary, Luc Lavoie, and sent him to Jakarta. According to one Barrick employee who was also on the scene, Lavoie was ineffectual. "He knew nothing about mining, nothing about Indonesia." According to the source, Lavoie failed to visit Barrick's office at the Cilandak Commercial Estate or to pick up a number of seminal books about Indonesian culture and politics that had been set aside for him.

Lavoie's lack of sensitivity became obvious when he and Vincent Borg, Barrick's director of corporate communications, invited a dozen Indonesian reporters to visit the company's large gold-mining operations near Elko, Nevada. "It was just gross," says one Barrick employee. "God knows how much that silly junket cost the company. It was the middle of Ramadan [the Muslim month of fasting] and here we were trying to wine and dine these unsuspecting fellows. Borg told me not to worry. He said he'd bought all kinds of

prayer mats and korans for the plane trip. He didn't have a clue
about protocol."

Once in Nevada, the jet-lagged Indonesians were brought up to
speed on Barrick's wondrous achievements. Borg told them that if
Barrick won Busang, the company would immediately set to work
building a US$1.5-billion mine. "We're certain this will bring sub-
stantial economic, social and community benefits to all the people
of Indonesia," Borg told them. "We're committed to protecting the
environment. And we won't seek loans. We have enough equity."
Placer Dome, the Indonesian reporters were told, was having financial problems. Its commitment to the environment was suspect.

Barrick officials do not deny casting doubts about Placer Dome.
They claim that the Vancouver-based company invited a discussion
of its record by claiming that it, not Barrick, was best qualified to
develop Busang. Fair enough. But there is evidence that Barrick went
even further, compiling a list of problems Placer Dome encountered
in Papua New Guinea, Venezuela, the Philippines and Kazakhstan.
A document bearing a fax number from Barrick's Toronto office was
transmitted to various media in Jakarta. It alleged that Placer Dome's
relations with various governments "have been poor to hostile."
Barrick officials admit that the information "may have been lifted
from a Barrick document," but insist that the company did not fax it
to journalists. Apparently, someone else did.

Barrick executives weren't the only ones putting their spin on events.
In mid-January, a few days before I was scheduled to return to
Canada, Yusuf Merukh called my hotel. With all the political and
corporate intrigue surrounding Busang, Merukh, the consummate
pest, had practically been forgotten. He had retreated to his home in
Perth, in Western Australia, but he was still pressing his claim for a
bigger slice of Busang. Now he wanted back into the limelight.
"There's no way I will back off," he told me. "I have some very
damning information that I think you should see. Why don't you
come down for a meeting?"

I landed in Perth late at night and was met at the airport by Merukh's driver, Trevor James. He ushered me into a large, late-model black Mercedes, and we set off towards my hotel. I asked James how long he'd been working for Merukh. "About six months," he told me. "It's been quite an experience. Mr. Merukh is rather strange." He didn't fill me in. It didn't matter; I would soon find out what he meant.

The next morning, James drove me to a shady residential district near the city centre. Merukh was temporarily working out of an office belonging to an Australian businessman named Warren Beckwith. The two men had joined forces and were planning on filing a lawsuit over Busang. Merukh met me at the door. A short, balding man, with a gangly piece of skin hanging over his right eye, Merukh was certainly strange *looking*. He greeted me nervously and led me into a room, where he pulled out a few old hand-drawn maps of Busang, covered with illegible scribbles. These maps, Merukh said, proved that Bre-X knew the Southeast Zone contained gold well before the company ever laid claim to the property. "The area's potential was recognized in 1988," Merukh insisted. "Bre-X had confidential information to that effect. But they never shared it with me."

Since he was a partner with Bre-X in the original Central Zone, Merukh argued that he should have a slice of the Southeast Zone, too. What's more, he said, his company had an option to purchase another twenty per cent chunk of the original site. When I asked him to show me the contract, he paused, and then said that his lawyer had all his copies.

Merukh lectured me for another hour, and then we went out for lunch. Although it was blistering hot, Merukh ordered soup and chased it down with a glass of sweet port. "Tell me," he said. "What do people say about me, about me personally?" I just about choked on my water and then muttered something innocuous. He didn't seem to be listening. "You know," he interrupted, "I've been speaking to Tutut. She doesn't want to deal with Barrick any more. Any day now, I think she's going to reach out to me for help." Merukh made several more outrageous claims before lunch was over. He had a strange habit of rubbing the top of his head while he spoke.

We had one more meeting before I left. Merukh insisted on taking me to dinner with his three daughters. Just before entering the restaurant, he took me aside and shoved a thick roll of Australian bills at me. "For your expenses," he said. I told him I couldn't accept any money. "No, take it," he said. I refused. He barely talked to me the rest of the night. "He thinks he knows the whole story," Merukh told his daughters, pointing to me and snorting. "Maybe he knows half of it."

James drove me back to the airport the next morning. "So, what do you make of this mess over Busang?" he asked. I told him I thought his boss had an outside shot at getting a piece of the property. The truth was, I wasn't sure what to think, or whom to trust.

12

THE TRIUMVIRATE

I sort of thought more of the shareholders than I thought of myself and my family, which I wouldn't recommend anyone doing.
– David Walsh, January 1997

The deal we have done is clean. There is no money involved, no signature, there is nothing. It's a very clean deal. You see, the good guys always win.
– Mohamad (Bob) Hasan, February 1997

THE PRESIDENT WAS UNHAPPY. Two of his children were bickering over Busang; that was reality, that was all that mattered. The squabble was embarrassing and it gave him a headache. Sudjana's ill-considered association with Barrick had ruined everything, casting a shadow on the country's entire mining industry. Members of the Indonesian Mining Association were scrambling to explain why Sudjana had forced Bre-X into a one-sided deal with Barrick and the president's daughter. "This is just an isolated case," blustered Beni Wajhu, the

IMA's secretary general. But at least one Western securities analyst based in Jakarta was already planning to revise his assessment of the Indonesian mining industry. Tom Soulsby, head of research for ANZ Securities Ltd., told me over lunch at the Jakarta Stock Exchange that Sudjana's move was "an absolute disgrace. Suharto must be quaking over this one."

As if all that weren't enough, the nationalists were making hay over the issue. Suharto's family was — once again — being portrayed as a group of meddlesome bandits in the international press. In the old days, Suharto's wife would have sorted things out in private. "My wife is my closest companion and loyal helpmate," the president once said. "There is no one else. If there were, the Suharto household would be the scene of an open revolt." But Madame Tien was gone, and the crisis was in the president's hands. He turned to his friends for guidance.

In December 1996, Suharto called a meeting. Joining him at his cattle ranch outside Jakarta were two of his most trusted associates: Mohamad (Bob) Hasan, a billionaire timber baron, and Jim Bob Moffett, the plain-talkin' head of Freeport McMoRan. Moffett did not like what was happening. All this nonsense with Barrick and Sudjana threatened foreign investment, he warned. It was bound to have a negative effect on Freeport's own share price. The company's stock had been flat for months. Rioting at its giant gold mine in Irian Jaya had caused a brief shutdown earlier in the year. Despite the best efforts of Freeport's public relations department, investors were still worried about political and social upheaval in Indonesia.

Bob Hasan outlined a solution. Give the nationalists their due, he counselled. Use public opinion as leverage to derail the Barrick juggernaut. Reassure the people that reason has prevailed, that an open auction will take place. Encourage other companies such as Placer Dome to make offers. Listen to all the proposals. Then announce that a joint-venture agreement has been reached among Bre-X, Freeport and the Nusamba Group.

One of Suharto's most powerful financial instruments, Nusamba is controlled by three of the president's *yayasans*. Hasan, who owns ten per cent of Nusamba, acts as its chairman and managing director. Under his direction, Nusamba — quietly referred to as the president's "retirement fund" — has grown into a US$5-billion empire, with interests in dozens of industries. No Indonesian would dare openly criticize a deal that involved the presidential conglomerate, Hasan reasoned. Barrick, he argued, wouldn't complain about being thrown out into the cold, because the company still had a group of COW applications awaiting presidential approval. If Munk wanted his company to have any kind of future in Indonesia, he would have to accept the decision.

Of course, Hasan stood to benefit. The diminutive sixty-six-year-old entrepreneur is a master of self-promotion. He has always used his long-standing friendship with Suharto to gain international respect — he is an executive member of the International Olympic Committee — and to amass a huge personal fortune. *The Economist* magazine estimates his personal wealth is at least $1 billion. "Mr. Hasan insists he is independent," *The Economist* added, sceptically. "But admiration for this charming fixer is tempered by concern about the greedy family he has served."

Hasan and Suharto met in the mid-1950s, when the future president was in command of a military post in Central Java. Hasan, a Muslim convert of Chinese descent, is reputed to have helped Suharto run a smuggling operation in the region. After seizing political power a decade later, Suharto returned the favour, handing Hasan millions of hectares of timber concessions in Kalimantan. "Through his control of the Indonesian Plywood Association, the Indonesian Sawmillers Association, the Indonesian Rattan Association, and the umbrella Indonesian Forestry Community, Hasan wields considerably more influence over the forestry sector than the Ministry of Forestry," writes journalist Adam Schwartz in *A Nation in Waiting*.

Hasan's other business interests include a bank, a national airline and a shipping company. He also owns *Gatra* magazine, one of the two weekly publications which gleefully reported on Sudjana's misappropriation of funds from the state-owned coal company. *Gatra's*

December exposé was the first salvo fired at the Barrick–Sudjana alliance, the beginning of a privately sanctioned attempt to drive Busang into the waiting arms of a Suharto–Hasan–Freeport triumvirate. It would not fail.

Placer Dome's CEO John Willson honestly felt he was being given a shot at Busang. The secret trip he made to Jakarta in December gave him "quite a strong feeling" that his company was in the running. He had no idea he was being used.

"There's a good chance [the file] will be reopened," Willson told me, a couple of weeks after returning from Indonesia. He'd met with some people "to see what the climate was," he added, and presented a proposal that went right to the president. Willson put himself on the side of the angels. Barrick's finagling had shocked him, he said. "My first reaction was horror. Bre-X persevered a long time down in Kalimantan only to see Barrick move in and change all the rules. Holy God, you spend money on exploration and you hit the big time, and then the rules change. Golly, who's going to take that bet? I thought Bre-X had a natural right to determine its own partner. Down in Indonesia, everyone I talked to agreed with me."

Norman Keevil, president of Vancouver-based Teck Corporation, also made a trip to Jakarta to meet with Hasan. Keevil left a few days later feeling confident that his company was also being considered as a potential partner in Busang. Hasan encouraged it; he told *The Globe and Mail* that he "liked Mr. Keevil's low-key approach and was impressed with Teck's mining technology." Soon reports appeared in the Indonesian press, suggesting that two state-owned mining companies had thrown their hats in the ring as well.

It was all a ruse. In mid-January, Nusamba eliminated a potential obstacle, snapping up a fifty per cent interest in Haji Syakerani's two companies, PT Askatindo, Bre-X's minority partner in the Southeast Zone, and PT Amsya Lyna, partner in another claim area to the north. This was the public's first clue that something was up; until that point, no one regarded Hasan and Suharto as active participants in the Busang sweepstakes.

Sudjana was directed to make an about-face. "The government encourages business people such as Mohammad Hasan to join the battle for ownership of the Busang gold project," he told the Indonesian press. "I don't want Indonesia losing out." With Nusamba sticking its foot in the door, there was no longer any danger of that. The triumvirate's plan was proceeding nicely.

Barrick pretended otherwise, but its bid for Busang was finished. Officially, the company had until 15 February to reach a final agreement with Bre-X and its minority partners. Nusamba's stake in Askatindo and Amsya Lyna signalled that this wasn't going to happen. Two weeks later, Nusamba made another telling move, acquiring majority control of Indocopper Investma Corporation, an Indonesian company with a 9.36 per cent share of Freeport's enormous gold mine. The unexpected deal — again brokered by Hasan — cost Nusamba US$312 million, with Freeport generously agreeing to backstop a loan, guaranteeing eighty-two per cent of any required financing. Freeport also pledged to lend Nusamba funds to pay interest on its new debt. It was a sweetheart deal and did nothing to dispel the perception that Suharto had Freeport in his back pocket.

On 6 February, Moffett landed in Jakarta. His presence didn't surprise anyone; he always flew to Indonesia at the end of Ramadan to pay his respects to the president and senior government officials. This year, however, his agenda was heavier than usual. "Jim Bob stayed a few extra days," said a Freeport official based in Jakarta. "I got pulled into the office and was told we were getting involved with Bre-X. I couldn't believe it. I mean, why would we do that? We already had our hands full in Irian Jaya. But the Indonesians were pushing hard. The first family was interested in getting its hands on everything it could." (The same official confided that it gave him great pleasure to see Sudjana's arrangement with Barrick thwarted. "Sudjana hates us," he told me. "He always wants us to play fast and loose, but we don't go along. He's tried to squeeze money out of us, any way he can. He's asked us for money flat out.")

The deal was announced eleven days later, catching most analysts in North America by surprise. Just after midnight on 17 February, Bre-X faxed a press release to various media, investment banks and investors, summarizing the arrangement. A new joint venture would be created to develop Busang. Bre-X would assume a forty-five per cent share in Busang Indonesian Gold JV; Askatindo and Amsya Lyna would get thirty per cent, the Indonesian government would get ten per cent, and Freeport the remaining fifteen per cent. Freeport would commit itself to providing US$400 million, or twenty-five per cent, of the estimated cost of construction, in exchange for the right to be sole operator of the mine. The company would also arrange up to US$1.2 billion in additional funding from Chase Manhattan Bank.

Some observers rejoiced that Barrick had been cast aside. "I'm just ecstatic that Barrick is out," Bre-X investor Greg Chorny told me. For his part, Peter Munk refused to confront reality — that he gambled Barrick's future by making a conscious decision to play footsie with Indonesia's ruling elite. It was sad, but hardly surprising. Great leaders seldom admit to making serious mistakes. Their goal is power; once this is achieved, their purpose is maintaining it.

"We've acted totally above board," Munk told reporters. The race for Busang, he added, "has become a game of Mr. Hasan's, a political game [which threatened to] transfer the destiny of our funds and our human resources . . . outside of our control. And that is contrary to the founding principles of what Barrick is all about. I think we all feel that this was going to be a bit too messy for us."

Time was running out on Busang, the big lie. Freeport was already making arrangements to send a team of geologists to the property. David Potter, Freeport's vice-president for exploration, would spend two weeks twinning Bre-X's best holes and analysing the core. By mid-March, Potter would have concluded that Busang was a fake, that the deposit had been recklessly salted. But it would take much longer to convince the rest of the world, thanks to a desperate public relations campaign by David Walsh, John Felderhof and some buddies on Bay Street. Many lives were about to be ruined.

It had not gone unnoticed that Bre-X had just lost half of its principal asset, which Walsh valued at more than US$20 billion. There'd been no mention of price. How much money would Bre-X actually get in exchange? Nothing. It was an expropriation. "I'd say that Bre-X just gave up a big chunk of Busang and got nothing for it," sputtered Ted Carter, David Walsh's old drinking buddy back in Calgary.

Curiously, Carter was one of the few market watchers to make that observation. Most analysts continued to back Bre-X, buying the hopeful spin that President Suharto's involvement meant that the development of Busang would proceed, directly and without delay. There were whispers that Yusuf Merukh's bothersome complaint — legitimate or not — would be settled internally. Approval of sixty-eight other Contract of Work applications, the majority from Canadian exploration companies, could finally proceed now that the Busang issue had been settled. Surely that was reason to cheer the new deal.

Canadian promoters backing the other junior companies were relieved that the Busang saga appeared to be winding down, but the whole affair left them wary. "It's been disconcerting for those of us who've been sitting on the sidelines, waiting for our COWs to be signed," said Robert Chase, president of South Pacific Resources Corp., a Vancouver-based outfit with five copper and gold prospects in Kalimantan. "I think we may see a number of the juniors reconsider their commitment to Indonesia, given what's transpired. I mean, what happens if someone else ends up with a big discovery? Will that be taken away, too?"

The analysts were singing another tune. Chad Williams, the boosterish gold analyst from Research Capital Group who had visited Busang the previous summer, told *Canadian Business* that, "given the circumstances," Bre-X had "got the best possible deal for its shareholders." John Ing, from Maison Placements Canada

Inc., told the magazine that "half a loaf is better than no loaf. I think Bre-X is very fortunate to get this."

Walsh unloaded some more ammunition, releasing yet another "updated resource calculation completed by KILBORN SNC LAVALIN. The current resource in all categories, as calculated by Kilborn, stands at 70.95 **million ounces of gold.**" This was a staggering increase of almost fourteen million ounces over the last estimate. It was unforeseen by everyone on the Street, even Bianchini. But the market was skittish. Bre-X shares listed on the Toronto Stock Exchange dropped $3 in two days, closing at $20.80 on 18 February. The company saw its market value drop approximately $400 million. Several major institutions, including Fidelity Investments, dumped most of their shares on the market. Why hang on, if half of Bre-X's assets had been taken away?

Bre-X had to address this issue head-on. On 20 February, Walsh hosted a crucial conference call with North America's leading gold analysts, including Egizio Bianchini from Nesbitt Burns and a host of other inveterate Bre-X boosters. Walsh was joined by John Felderhof, Rolly Francisco and two representatives from J.P. Morgan, Leslie Morrison and Doug McIntosh. Their task was to convince the analysts that Bre-X's true value had not been diminished at all, that Bre-X had been offered an excellent deal. Their audience adored them; this mission was a cakewalk.

"The success of these negotiations is a major victory for Bre-X," Walsh declared at the outset. "The program is exactly as we had always envisioned it. Unfortunately, I don't think that the market understands the deal. It's clear from yesterday's trading that there is a misconception by investors and analysts about how big a win this is for Bre-X."

Walsh claimed that Bre-X had never really controlled ninety per cent of Busang. "This was never the practical reality," he insisted. "Nor was it ever the basis for the valuation of Bre-X's stock." But this contradicted a statement by Leslie Morrison, the company's economic advisor from J.P. Morgan. "We were retained by Bre-X on the third of September, 1996," he noted during the same conference call, "and we prepared a valuation at that time, assuming that they did, in fact, own ninety per cent."

That wasn't the only contradiction. "It was always envisioned that we would share in the ownership with local interests and a development partner," Walsh asserted, failing to mention it was always envisioned that Bre-X would receive cash for a share in the ownership. Nine months earlier, he had told his shareholders that "we'd want a walk-in, cash payment for twenty-five per cent of Busang. I'd say $2 billion for the twenty-five per cent sounds about right."

Walsh's blathering was nothing compared to Felderhof's indiscretions. "We are dealing with a very unusual deposit here and it's one of a kind in the world," he said. "Once we finish the drilling program we are currently doing...my estimate is ninety-five million ounces. Mike de Guzman, my project manager, he estimates 100 million ounces. If you would ask me what is the total potential, I would feel very comfortable with 200 million ounces. So far as I am concerned, there's lots of blue sky there." It was incredible, and yet none of the analysts flinched.

Then Bianchini lobbed a marshmallow at a flustered Morrison, who took a big swing and missed.

BIANCHINI: Could you say that this is the best deal, the most fair deal, if I can use that term, that Bre-X has seen? That includes the Barrick deal or any other deal that was proposed.

MORRISON: It is certainly the case in the real world that this is the only deal which is acceptable, which is available to Bre-X. This is the one which was available and it's the one which has been blessed by our local partners and I think it's true to say it's the one that the management of Bre-X and its advisors is [sic] particularly pleased about in its totality, this is simplicity, the size of its direct

ownership, the way in which it's been welcomed
by the Indonesian partners and the government

A little later, Bianchini had another odd — but revealing — dialogue, this time with Felderhof.

FELDERHOF: Egizio, can I just add one point? I think what happened here is that we found too much gold, so it therefore becomes a question of national interest. If we had found a lot less, this [ownership dispute] would never have happened.

BIANCHINI: I totally agree with you, John, but again, I think the shareholders, you know how the shareholders feel, John, you're one of them and I think we're going to have to look at Indonesia in a much different light than they have presented themselves at conferences around the world on their road shows. I think all of you out there would agree with me that you have to look at it a lot differently than it used to be.

FELDERHOF: Yes, I'm a bit disappointed. But you've also got to appreciate this is a very unique situation and even with forty-five per cent I feel that Bre-X could end up with ninety million ounces. That's my feeling and that's a pretty good mine.

BIANCHINI: From your lips to God's ears, John. Thank you.

Felderhof then went on a rampage. His 200-million-ounce figure was "conservative." Some of his new holes were hitting thirty to eighty grams of gold per tonne, he said. He envisioned an open-pit mine at least six hundred metres deep, six kilometres long and three kilometres wide. It would pay for itself inside two years. The gold was at least ninety per cent recoverable, he claimed, well above the industry average. To top it all off, there weren't any pesky Dayak digging and panning for gold in the area. "A lot of other companies experience, in Indonesia, the illegal mining problem," he said. "At Busang, we have no illegal mining operations, no influx. It is very unique." Actually, it was incomprehensible. If Busang really hosted the world's largest gold deposit, the Dayak would certainly know about it.

No one questioned Felderhof. As was their habit, the analysts didn't *analyse* the geologist's claims; they merely repeated them as though they were fact. Kerry Smith, an analyst with First Marathon Securities Ltd., issued a report immediately following the conference call, rating Bre-X a "speculative buy" at $21.50. "*The Busang deposit will continue to grow*," he advised. CIBC Wood Gundy's Bruno Kaiser quickly followed. "We share in the optimism of a 200 million oz. (or more) reserve since a property visit we had last summer. . . . We are raising our ranking from Hold to Buy."

Sensing that the market had turned irrevocably in his favour, Walsh began to swagger. He fired off a long letter to shareholders, stating that "modern-day 'claim-jumpers'" had tried to horn in on Busang, a gold field of "extraordinary riches." Walsh wrote that the "increasing value for the property also exacerbated the difficulties in negotiating with the Republic of Indonesia. In a word, we were ultimately victimized by our own success." But he had prevailed. "The alliances we have forged, together with the majority interest we retain in this unprecedented resource, will set the stage for Bre-X to become a driving force in the gold industry well into the next century."

On 12 March, Walsh called me from the Royal York Hotel. He was in Toronto with the entire Bre-X gang, including Felderhof, de Guzman and the Filipino geologists, attending the Prospectors and Developers Association convention. Walsh and Felderhof were the talk of the town, the star attractions. Bre-X investors had made pilgrimages to Walsh's suite, asking for autographs. *The Northern Miner* had thrown a dinner party in their honour at a local restaurant, presenting them with a pair of Mining Men of the Year trophies. And Felderhof had just been presented with the PDA's Prospector of the Year award, launching a thinly veiled attack on Barrick in his acceptance speech.

Looking sweaty and nervous and drunk, Felderhof started off by mumbling a tribute to twelve former colleagues, "lost in helicopter crashes, land slips, and flash floods." Then he castigated the "tree-shakers" who'd tried to steal Busang away. "When you find something big, somebody else likes to take control of it," Felderhof slurred, clutching the microphone stand. "There are the companies that shake the tree and see if something will drop out. The only exploration they do, when they explore, they explore the ways and means of how to get something from you. The jungle they know is the concrete jungle, all right? The closest they get to rocks is scotch on the rocks. And I say to them, go and find your own."

Felderhof waved his arm in the general direction of Mike de Guzman, who was standing at the side of the room, away from the lights. "I cannot say enough about the team I had," he said. "O.J. Simpson had his team, his dream team. I got mine. Mike de Guzman, my exploration manager, I think he's hiding in the corner somewhere over there. Uh, Cesar Puspos...Jonathan Nassey, they're just here among many others. I'm just proud of them, and I'm sure they're proud of me."

Later on in the evening, when the PDA named Peter Munk as Developer of the Year, a group of Bre-X employees got up from their seats and left the room. "It was great," Walsh told me during our conversation two days later. "We really made a point right there." He went on about a new mining company he was starting in Armenia, an outfit called Arm-X. "We're going public in June," he said. "My brother-in-law is going to be president." I asked him if Felderhof would be involved in exploration. Walsh's voice dropped.

"Maybe."

What about de Guzman?

"No," he said. "I don't think so."

Walsh had a secret. Jim Bob Moffett had called from New Orleans, just a few hours earlier. "David," he said, "we've got a problem. You need to get some of your people back [to Busang]." So de Guzman was dispatched. It would be his final assignment.

Jim Bob's boys had their doubts before they even arrived at Busang. There were plenty of red flags flying over Kalimantan. Harkening back to concerns raised among industry insiders a year earlier, the men from Freeport worried that there'd been no independent evaluation of the property, no data to indicate that the fabled Southeast Zone contained any gold at all. Freeport had asked for some core samples and didn't get any. There were none. Bre-X hadn't split any of its core, and that wasn't "standard operating procedure," Moffett told journalists during a conference call, two months later. The area was tightly secured by Bre-X's Filipino managers. "Very few people other than [those on] guided tours had ever been at this job site and given any ability to move around," Moffett said. All those analysts and their solemn predictions and guarantees meant nothing to Freeport. "We weren't caught up in any hype. We were starting from scratch."

Freeport started with the appropriate premise: Busang defied logic. Investors and analysts and even Peter Munk had been content to believe in the story, but Freeport wanted hard evidence. It operated the world's biggest gold mine, with seventy million ounces. All that gold didn't appear overnight. The company had spent almost two decades working a copper mine in Irian Jaya before finally hitting gold a few hundred metres to the north. Walsh and Felderhof and de Guzman — three slobs Moffett had never heard of — were bragging that they had beaten his record, thanks to some ill-defined theories, murky "reinterpretations" and "perceptual thinking."

"We can take gold out of the ground once we find it, but I have not been ordained to put gold into the ground," Moffett said, in his good ol'-boy drawl. "Let's face it. These people were reporting that this was the largest gold deposit since our Grasberg deposit, and we know what it takes to find a Grasberg. So we had to be sceptical... and we went from being sceptical to being cynical."

Freeport's team of geological fact-checkers had arrived at Busang on 1 March. De Guzman was there to meet them, and he spent the next couple of days helping Freeport assemble its drilling rigs next to the "sweet spots" — areas drilled by Bre-X that supposedly showed extraordinarily high gold readings, some with high grades right at the surface.

Problem: "There was no gold at the surface," said Moffett. "While we were preparing the drilling rigs, it [didn't] take long to take grab-samples from the topography that's exposed. Generally, gold deposits that are this close to the surface have some geochem halos, or geochem surface anomalies. We could find none. That was a big concern to us."

It made sense to compare notes with de Guzman. Except by then, he'd flown the coop, taking Cesar Puspos, Jerry Alo and Jonathan Nassey with him. Moffett was incredulous. His company was attempting to outline what the Bre-X geologists claimed was a unique ore body. A multibillion-dollar deal hung in the balance. And yet no one in any position of authority from Bre-X was there to lend any insight. After three years of vigilantly guarding Busang from outsiders, not a single senior manager remained behind to keep an eye on this band of strangers.

"They had all left the job site," Moffett said. "[They were] in Toronto at this big affair." His crew carried on with its work. The intention was to collect a small but highly representative group of drill core, have it assayed and compare the results to Bre-X's remarkable findings, all of which had, apparently, been endorsed by Kilborn Engineering.

While Bre-X played at the PDA convention, Freeport drilled seven holes inside a five-square-kilometre patch of land within the

Southeast Zone. Four "twin" holes, all around 250 metres deep, were punched parallel to four very compelling Bre-X holes showing high grades of gold. Two of those Bre-X holes — 63 and 202 — showed high grades of gold starting right at the surface and continuing for more than 250 metres. At some intervals, the gold reading was higher than ten grams per tonne. Three more "scissor" holes were drilled at cross angles of 60 degrees, intersecting a total of fifteen other hot Bre-X holes. Had Bre-X's own hole-by-hole readings been genuine, there was no chance Freeport could miss hitting gold.

The drilling took about a week. More than 1,750 metres of fresh core were pulled from the ground. The samples were split, bagged and shipped to four different labs for assaying: Indo Assay in Balikpapan, Inchcape Testing Services and Sucofindo, both in Jakarta, and Freeport subsidiary Crescent Technology Inc. in New Orleans. Two different methods were used — the standard fire assay, and the cyanide leach method that Bre-X preferred. Freeport also tested four samples of solid core that Bre-X had drilled and left behind in plastic bags at Busang. In addition, Freeport grabbed five hundred samples of crushed core from Bre-X's warehouse in Samarinda.

The results were, of course, disastrous. Freeport found no more than a few *one-hundredths* of a gram of gold in any of its samples, about the same volume contained in ordinary sea water. *Hole 63 — Negative. Hole 202 — Negative. Hole 173 — Negative.* It went on and on. The solid core Bre-X had drilled and left lying around Busang was barren, too. The only gold bits of any note came from Bre-X's crushed core at Samarinda. A metallurgist had discovered that the gold grains in these samples were uncommonly large — up to two millimetres thick — and featured a high content of silver in the centre. This is typical of alluvial gold, found in rivers and streams, not igneous gold found in bedrock.

These results told Moffett just about everything he needed to know: Busang was a wasteland, the ore had probably been salted, the Bre-X boys were phoneys. In his mind, this was "just a real ugly case of desperate people that needed to find a gold mine and couldn't find it. So they invented one. We knew that our [reading of] insignificant gold was going to be pretty prophetic," Moffett added. "The problem is, no one wanted to believe a thing like this could happen."

He didn't come right out and say that, at least not right away. When Jim Bob called Walsh after getting the news, he had just received one set of results from Indo Assay, based on the first three holes Freeport drilled. "I hated to interrupt their joy," he recalled. "I just wanted [Walsh] to be aware that things weren't going normal. . . . Obviously when you got a guy sitting there with all the publicity, and we're here, and we've been on the job site for two weeks and I'm fixing to call him and tell him that we've got three core holes, and he's got three hundred, and that there's insignificant gold . . . I didn't want to impose on him and try to explain something which we were really just trying to sort out ourselves." What he did say was that Freeport had encountered "a problem, and it's a pretty significant problem, because our assays don't check your assays."

Walsh reacted oddly. He didn't sound amazed, or disbelieving, or even mildly outraged at Moffett's shocking revelation. Instead, he offered what Bre-X investors would later recognize was his pat response when confronted with evidence that Busang was a fake. "He started out by telling me he was a financial person and he didn't understand all that stuff," said Moffett. "So I was talking to a guy who was telling me that he really only understood the financial side of the business. [Walsh said], 'I'll have to get Felderhof involved.'"

Felderhof was only slightly agitated when he heard the news. His explanation, which he would use again and again, was that Freeport had made a mistake. There must be some confusion, he told Moffett. You must have mixed up the drill holes. "I said no, that's not the case, because we have resurveyed those locations. This is something that's much more germane to the technological part of this project than confusion." Moffett told Walsh and Felderhof to send someone back to the job site. As Bre-X's senior vice-president of exploration, it would have been reasonable for Felderhof to make the trip. But he didn't. He dispatched de Guzman instead.

"We waited, and waited, and waited," Moffett said. "And he never arrived."

13

LOOKING FOR BIGDUDE

Lads, get a grip, this is serious. The key BXM geologist on the ground in Indonesia has disappeared without a trace as Freeport is conducting due diligence on the site. Connect the dots dudes. Now is the time to be extremely cautious and ask the tough questions. . . . Man, I hate this crap. I can't sleep, I can't concentrate on anything else, I am in denial that my friend Mike de Guzman is gone, I am a mess.
  – Internet postings by Bay Street analyst Bill Stanley, a.k.a. Bigdude, March 1997

IT'S TEMPTING to believe David Walsh's story, the one his private investigators whipped up for him, funding provided by Bre-X. Walsh wants the world to believe that de Guzman committed suicide by jumping out of a helicopter en route to Busang. "He killed himself," Walsh's $1.25-million report states, "when faced with the prospect of having to be exposed for salting Bre-X samples." Very good. Nice and simple. Does not incriminate others.

This might seem like a tidy solution, but it's not. Michael de Guzman was a thief, a polygamist, an inveterate liar. Not stupid. He knew the Busang scam was unravelling. The truth had to emerge with the arrival of Freeport's geologists. He hadn't panicked in the run-up; de Guzman was with Walsh and Felderhof at the PDA. He was beaming, according to witnesses, and flashing pictures at the big awards banquet. He visited Niagara Falls with his pals Cesar Puspos, Jonathan Nassey, John Salamat and Jerry Alo. They went shopping and hung around a Toronto strip joint called For Your Eyes Only. De Guzman fell for one of the dancers there, bought her gold jewellery, regaled her with stories about his tremendous wealth.

Why would he fly halfway around the world, back to the scene of the crime, and dive out of a helicopter?

This isn't the first theory that Walsh has floated. The initial explanation came right after de Guzman "jumped," when Walsh still had "absolute confidence in the integrity and accuracy" of Busang's one-sided technical data. De Guzman, he said, could no longer face living with hepatitis B. The Bre-X investment relations team, fresh out of the University of Calgary, told telephone callers that the geologist's hepatitis B was "at the terminal stage."

When de Guzman got the call to return to Busang, he brought Nassey, Salamat and Alo with him on the first leg of the eighteen-hour journey. They flew back via Hong Kong; de Guzman didn't know it, but the man sitting next to him in the first-class cabin had also been at the PDA convention and was "a friend" of a Freeport executive. The passenger says that Alo and Salamat moved up from business class and had a brief argument with de Guzman over money. All three had made millions cashing Bre-X employee options on the open market: de Guzman, $4.6 million; Alo, $1.2 million; Salamat, a bit less. They wanted to know why de Guzman had so much more.

When the party arrived in Hong Kong, Alo, Salamat and Nassey caught a four-hour connection to Jakarta. De Guzman, meanwhile, remained at the Hong Kong airport for several hours before flying on to Singapore, where he had a scheduled medical appointment at the Mount Elizabeth Hospital. He spent two days in a Singapore hotel with his wife Gini, from West Java, who'd flown up to meet him. On 17 March, de Guzman arrived at the hospital for his check-up. This included a cardiac stress test, which de Guzman passed easily. He faxed the results to his sister Diane, a nurse in Los Angeles. Doctors had given her brother "a clean bill of health, with no limitations whatsoever." De Guzman's hepatitis B was clearing up.

He left Singapore the next morning and flew to Jakarta, then made the four-hour flight to Balikpapan. He checked into the posh Altea Benakutai hotel and called Lilis, his wife in Samarinda, telling her that he was going directly to Busang and that he'd see her in a few days. De Guzman stayed up late, drinking with a subordinate, Rudi Vega. Vega claims the two men ended up in a karaoke bar at the Balikpapan Hotel, drinking beer. At one point, de Guzman warbled out his favourite Frank Sinatra tune, "My Way." A few hours later, he was gone.

Peter Von Veen is the general manager of PT Indonesia Air Transport (IAT), the company that leased Bre-X its Alouette helicopter. A burly ex-airline pilot, Von Veen was born in a Dutch internment camp in Indonesia during the Second World War and has lived in the country almost his entire life. He invited me to his comfortable house in south Jakarta to "set the record straight" about de Guzman, two months after the Bre-X geologist disappeared. Von Veen had been surfing the Internet, tracking down what was being written about de Guzman's apparent death; it seemed that some people believed that IAT had something to do with it. "All this speculation, it's gone from the bizarre to the ridiculous," he told me.

Von Veen says that at nine o'clock the morning of 18 March, de Guzman and Vega showed up at the Balikpapan airport. They were over an hour late, which was unusual. IAT had been flying Bre-X in and out of Busang for three years, and de Guzman was normally very punctual. He was wearing shorts, a T-shirt and a denim jacket, and was carrying a vinyl handbag and a black leather briefcase.

There are four doors on an Alouette, two in the front and two sliding doors in the back. Inside, there's room for six passengers; de Guzman and Vega settled on the bench-seat in the rear. They were both very familiar with the aircraft. Bre-X had chartered that particular model for three years, at a base rate of US$38,000 a month. The flight crew was never exactly the same, however; IAT has twenty helicopter pilots in its stable, most of whom are "semi-retired" military flyers over the age of fifty-five. On this particular occasion, the pilot was Edy Tursono, a seasoned veteran who'd been with IAT since 1992 and had flown to Busang at least five times before. He was the pilot who'd flown the group of Canadian analysts — including Bill Stanley — to Busang the previous summer. Tursono piloted from the right side of the helicopter. To his left sat flight engineer Andrian Mailan, with eleven years of active duty.

It takes just over one hour to fly from Balikpapan to Busang; however, IAT preferred to stop and refuel at Samarinda, which was at the halfway point. Both de Guzman and Vega were manifested to go straight through to Busang after stopping at Samarinda.

According to Edy Tursono, that's not what happened. De Guzman and Vega were met at the Samarinda airport by a pair of Bre-X staffers, one male, the other female. They had a brief conversation and appeared to leave the airport. When de Guzman returned, he was wearing full-length jeans. He was no longer carrying the black leather briefcase. De Guzman told Tursono that Vega was staying behind, and he climbed back into the rear of the helicopter.

Both doors were shut properly, and pilot, engineer and passenger were strapped in. Ground control gave clearance, and the Alouette flew off in the direction of Busang. Visibility was good; Tursono manoeuvred the chopper to 245 metres, just under the perennial layer of cloud cover. All three men were wearing earphones; without them, they would have been deafened by engine noise.

De Guzman allegedly bailed out about twenty minutes into the flight. Tursono and Andrian hadn't noticed him undo his seat belt and remove his earphones. Their story is that he grabbed the handle on the left door and pulled it towards him. The door flew open with a bang and the helicopter jolted upwards. As Tursono struggled to maintain control of the aircraft, Andrian looked back. De Guzman was gone. "It was damn considerate of him to jump from the left side," says Von Veen. "Had he gone from the right, he would have struck the rotors and brought the chopper down."

Tursono reduced speed and punched a button on the satellite navigational device in front of him, recording the current latitude and longitude positions. He called back to base in Balikpapan, gave a brief rundown of events, and began a low-level search over the area where de Guzman had supposedly jumped. Because the swampy ground below was covered by a thick jungle canopy, Tursono and Andrian could not determine where de Guzman had landed. After twenty-five minutes, Tursono was instructed to head back to Samarinda and report to police.

Four days later, a ravaged and decomposed corpse was pulled from the jungle, just a hundred metres from the spot indicated on the helicopter's tracking device. Family members say the fragmentary remains indicated the corpse belonged to de Guzman. Autopsy reports from Indonesia and the Philippines seem inconclusive.

Indonesian authorities say that de Guzman left a number of items behind in the helicopter, including the vinyl handbag. Inside was a Rolex watch, a wallet, some gold rings and the infamous "suicide note," which Walsh immediately used to explain de Guzman's abrupt departure. Von Veen showed me a photocopy of the suicide note and allowed me to compare the handwriting to several IAT forms de Guzman had filled out. The handwriting appeared to match, although the suicide note mostly featured large, capital letters. The message is inscrutable. Scrawled across six pages of lined paper, the note reads:

Authorization Letter Full Authority given to Mr Bernhard Leode [Bre-X's Jakarta-based financial controller] *to represent act on my behalf and for my behalf in case of disability or my death. Voluntarily issued.*

My final request to Rudy — rm 914 Pls bring my black bag w/all my very important notes must hand carry to office and Bogor key here thanks Mike

Call 02115228253 [telephone number for Bre-X's Jakarta office]
Bernhard Leode *Pls accompany my body (death) to Manila document for my wife Teresa Pls hand carry including my passport
*In Jkt Do not bring my body to Bogor. Stay at Funeral Parlour while waiting for Travel to Manila.

Settle accounts. Personal.
Thank you very much my dear friend.
Mike de Guzman

God bless you all No more stomach pains!! No more back pains!! To Bernard for my wife Teresa. My request do not bury me burn — cremate me in Manila

To: Mr. John Felderhof + all my friends sorry I have to leave I cannot think of myself a carrier of hepatitis "B" I cannot jeopardise your lifes, same w/ my loved ones

Several more notes turned up, apparently procured from de Guzman's black bag. Addressed to his first wife and children in the Philippines, they refer to the "sickness" that "is killing me." Please, de Guzman wrote, "be strong and continue. . . . I love all of you. . . . I die without regrets."

Lilis de Guzman did not receive a final message. She last heard from Michael the night before he disappeared, when he called to say he'd see her soon. Sitting alone in her shiny new home in Samarinda, the one Michael bought, she ponders over a card he sent from Toronto. "I'm thinking of you every moment I'm awake." Lilis didn't know Michael had three other wives. "I found out about the children in the newspapers," she says.

I wish I knew what happened to de Guzman, but I don't. Bre-X says he killed himself. His family thinks he was murdered. Others feel he may still be alive, living on some remote, guarded estate somewhere.

A fascinating, possibly gruesome mystery, but it doesn't really matter. Whatever the circumstances surrounding de Guzman's sudden departure, the fact that it came at such a critical moment should have tripped alarm bells up and down Bay Street. It was a very suspicious event; investors who ignored it were badly burned.

Bill Stanley was one of the lucky ones. He sold all his remaining Bre-X shares on the news that de Guzman had "fallen" from the helicopter. "It was too weird," he says. "The idea de Guzman had committed suicide didn't make any sense to me. I'd seen him at the PDA and far from being unhealthy, he was energetic and seemed very happy. Then I remembered some of the old stories I'd heard about salting, and geologists disappearing. I thought about New Cinch, and that assay lab worker who'd been murdered. I had so many questions about Bre-X since my trip to Busang that I figured it was time to bail out."

Stanley turned to his computer at work and logged onto the Internet. Within seconds he was connected to a freewheeling, loosely patrolled Bre-X chat group, a so-called thread that was part of a large, California-based network called Silicon Investor. The Bre-X thread was a clearing house for company-related gossip, newspaper articles, brokerage reports and market updates, and was heavily punctuated with lunatic patter and puerile personal attacks. As the Busang saga gained international prominence, the thread attracted dozens of contributors and many more lurkers, silent voyeurs. Monitoring the thread every day from his office in downtown Toronto, Stanley was appalled by the quality of some of the information being posted. He figured he could do better. By late February, he was sending in his own messages. His handle was "Bigdude."

Stanley hadn't come to any firm conclusions about Busang when he first started posting. He was, he says, simply sharing what he knew, which was far more than most posters. He'd been to Busang, after all. He knew how to read marketplace tremors. Unlike the gold analysts who were covering the company, he was much more detached.

He used a pseudonym because he didn't want his bosses to know he was posting. A branch of the investment bank where he worked had done some early business with Bre-X, and Stanley didn't want to get caught in any conflict. A lot of the Internet posters used fake names. There were Turk, Drumbeat, The Mouth, Goldman and Nasdaq. The most frequent contributors were relentlessly pro–Bre-X. They took umbrage at anyone who dared write negative messages about their beloved company. Bre-X bashers were slapped up and down, accused of either trying to short the company's stock by spreading damaging rumours, or working for Bre-X's many enemies.

On the flip side, it's alleged that Bre-X had operatives working the thread. An obsessive poster dubbed Mikesloan, for example, went to great lengths trying to convince readers that Bre-X was the target of a "huge clandestine operation.... I don't believe for a second that Bre-X is involved in any fraud ... Barrick Gold is behind all Bre-X's problems.... They are trying to destroy Bre-X.... Freeport has been set up by Barrick," et cetera. (Not surprisingly, Barrick officials tracking the thread paid particular attention to Mikesloan. They claim he was David Walsh's brother, Richard, also known as Merrick.)

The Internet has become a significant weapon in every penny stock company's PR arsenal. Practically anything can be "published" electronically, instantly and without attribution; some companies have succumbed to the temptation and "salted" the Net with phoney claims. Philip Lieberman, an elderly mining promoter, caused an uproar in early 1997 by posting messages that his company, American Technology Exploration Corp. (ATEC), had discovered seventy-six million ounces of gold in the Nevada desert. ATEC was delisted from the Vancouver Stock Exchange after Lieberman was found out, and the British Columbia Securities Commission banned him from trading on the VSE for life, concluding that "ATEC used the modern technology of the Internet for an old-fashioned purpose — promoting its shares with outrageous misrepresentations."

On Wednesday 19 March, Bigdude posted the first of many messages that cast serious doubt on Bre-X's credibility. "PEOPLE JUST

DON'T FALL OUT OF ALOUETTE HELICOPTERS!!!...Be careful dudes this is truly Bizarre even by my standards. There will be a logical explanation, but we may not like the answer." The post enraged the pro–Bre-X crowd. Bigdude was accused of sensationalism, of needlessly scaring off investors. But Bigdude continued to poke. "How is Freeport making out with its Due Diligence?" he asked. "Anyone heard about what kind of questions were going to be asked of de Guzman had he not disappeared?" This drew more flames from the Bre-X believers.

They should have paid attention. Bre-X shares, which had been trading below $20 since the Freeport announcement, began to free-fall the next day. A story from *Harian Ekonomi Neraca*, an Indonesian business daily, was picked up by an American wire service and then posted on the Bre-X thread. The article quoted two unnamed sources, one from Freeport, the other from the Indonesian Department of Mines, suggesting that Busang was a bust. "Possibly, the deposits are not as large as has been reported," read the translated copy. "In fact, it is possible that the deposits are not worth mining." More than ten million Bre-X shares changed hands in Toronto and New York. The stock closed at $14.25, down $2.25, shaving $800 million from Bre-X's market capitalization.

Bigdude kept posting over the weekend. "There appears to be a leak from the assay lab that Freeport is using in Jakarta for certain assaying. Scuttlebutt out of there is nasty." Others responded that the gold had to be there — Kilborn had said so. Think again, wrote Bigdude. "Kilborn did not actively perform due diligence, it was not in their mandate. . . . Remember, Kilborne [sic] worked for BXM and made a lot of money as a result."

Bre-X forcefully denied anything was amiss. Met by reporters on his way to Manila for de Guzman's funeral, Felderhof said he was "one hundred and ten per cent confident the gold is there. I'm getting tired of all these stories." Walsh fired off an angry press release, decrying the "continuing proliferation of falsehoods and misinformation based on unsubstantiated allegations by unnamed sources . . . . Unfortunately, when the first ounce of gold is poured at

Busang, I am sure the naysayers will complain about the colour."
He threatened to sue "certain parties and publications, due to the
erosion of share value which has resulted from these reports."
(Mikesloan posted that "maybe Bre-X's lawsuits should start right
here on the Bre-X thread.")

Felderhof and Walsh did not mention that Jim Bob Moffett had
told them it found "insignificant gold values" at Busang. They didn't
let on that their lawyers in Calgary had received a table of negative assay results from Freeport. Nor did they reveal they had already
dispatched three engineers from Strathcona Mineral Services to
Indonesia. Strathcona's mandate was "to undertake an independent
review of the technical work carried out by both Bre-X and Freeport
in order to determine the reasons for the differences in drilling
results from the two companies."

There clearly *was* a problem, but Walsh chose to keep quiet.
Investors — retail and institutional — were frantic with worry,
particularly those overweighted with Bre-X stock. They naturally looked for answers from the big-name gold analysts who'd
recommended the company. Of all the analysts, Egizio Bianchini
pulled most weight, and therefore faced the most pressure. Nesbitt
Burns was the largest seller of Bre-X stock. Bianchini was still the
most enthusiastic Bre-X analyst in the business. His reputation
was riding on Bre-X, not to mention his own equity. His friends
and family had bought in, thanks to his glowing recommendations. Now the bell tolled, and he had to make the biggest call of
his life.

He wasn't about to turn back. Bianchini came out bullish on
Monday, 24 March, advising clients to buy Bre-X, reiterating his top
"S-5" rating and setting a target price of $29, almost double the
stock's value that day. During a pair of extraordinary conference calls
he made with money managers in Canada and the United States,
Bianchini insisted that Busang was legitimate. "I suspect that
Freeport's initial results were probably quite low," he said. "Now, do
not be alarmed. I'm going to explain this.... The gold is there.
There's no question about it."

Bianchini tried to patch the burst bubble, but his reckoning was flimsy and contradictory. Busang's incredible drilling tests had "been looked at by their consultants and by all, by world experts," he said. "Three separate entities have sampled this deposit, and all three entities have come up with results similar to Bre-X. In some cases higher than Bre-X. And those are the only entities apart from Freeport and Bre-X that I know of that have actually done any work apart from the analysts that have visited the property.... So you know, every entity that has looked at this has, has basically walked away from it, you know, saying yeah, the gold is there."

Just who those "three entities" were is still a mystery. Bianchini could not say. He was "being vague for confidentiality reasons," but he virtually eliminated Kilborn from the list. Bianchini admitted that the vaunted engineering firm — which had lent so much weight to Bre-X's claims — "[had] not sat on every hole. They were supplied core from Bre-X.... They are not responsible for the drilling, they are not responsible for the work themselves.... I, I don't know, I can't tell you with one hundred per cent accuracy to what extent Kilborn has, shepherds the process."

As for the discrepancy in the assay results, Freeport could have used the wrong assaying procedure. "That's sort of what I'm, what I'm hearing.... If Freeport had indeed used fire assay as opposed to cyanide leach, guess what? They are not going to get accurate results." That didn't satisfy one astute money manager. Why, she asked, would Freeport use a method that was doomed to fail? "Again, we're just, we're all looking for, for possible answers. So, unfortunately, and I hate doing this, we have to speculate on what, what method they're using." Then he did another 180-degree turn and said that "one has to assume that they did not use fire assay." *The gold is there.*

What if Freeport had used the "proper" cyanide method, asked the same money manager. "My conclusion would not be different," Bianchini said. *The gold is there.*

He raised the ugly spectre of salting all by himself, and dismissed it. "Many of you out there have heard speculation that, you know, the lower than expected results for grade from this deposit are some-how linked to Mr. de Guzman's death. . . . Unfortunately, he is dead. And you know, any of you that know the man, that knew the man, I'm going to say publicly here, know that the possibility of any funny business on the part of Mike de Guzman or his knowledge of any funny business, I mean I just, it's, it's, it's just impossible for me to believe that. . . . If there was any funny business going on he would have been long gone long, long ago. So to those of you that, out there that are speculating that, that, you know, that the core was salted or that somehow the samples were salted or somehow that the numbers were jigged, I mean it is so preposterous I just, I just, I'm not even going to address the possibility of that happening because I just, it's infinitesimal."

Bianchini wrote a formal update for Nesbitt Burns. Bre-X posted it on its own corporate Web site, for everyone to read. "The recent reports coming from Indonesia indicating that the Busang deposit is smaller than indicated by the company and that it may not be eco-nomic are very likely erroneous. . . . The gold is there!"

Bianchini was setting big targets; his colleagues on Bay Street either stayed silent or heartily endorsed Bre-X. Kerry Smith, gold analyst at First Marathon, and Michael Fowler, at Levesque Beaubien, issued quick reports recommending Bre-X stock as a buy. Was it a coinci-dence that all three brokerages had underwritten Bre-X stock? In the few days following their forecasts, more than fifteen million Bre-X shares changed hands. The enormous volume indicated that opin-ion about the Busang deposit was deeply divided; investors either believed there was gold, or that there was nothing at all. By 26 March, investors were locked in. Trading had been halted; Bre-X was preparing an important announcement.

It was a bombshell. "Earlier today, Bre-X was advised by Strath-cona Mineral Services Limited, an independent Canadian mining industry consultant, that there appears to be a strong possibility that the potential gold resources on the Busang project in East

Kalimantan have been overstated because of invalid samples and assaying of those samples."

In case someone missed the meaning, Freeport issued its own release, formally stating that it had "drilled seven core holes within the [Southeast zone] to confirm the results of core holes previously drilled by Bre-X. To date, analyses of these cores, which remain incomplete, indicate insignificant amounts of gold."

The rumours had been correct after all, yet Walsh argued that there'd been a mistake. Strathcona would have to spend another month at Busang and conduct its own drilling, he said. The core would be tested every way possible. Bre-X, he said, would be vindicated. But mutual fund managers prepared for the worst, drastically writing down their Bre-X holdings. Most fund managers assigned Bre-X a share value of $1 to $3. This was a sure signal that Bre-X stock was going to plunge once trading resumed. It was also an act of self-preservation; if fund managers didn't act quickly and write down the value of their Bre-X holdings during the remaining cease-trade, then investors could sell their fund units at inflated values. Mainstream Canadian funds that had relatively high exposure to Bre-X — such as Greenline Precious Metals (8 per cent of its portfolio was tied up in Bre-X), CIBC Precious Metals (8.1 per cent), and First Canadian Precious Metals (4.83 per cent) — virtually wrote off their Bre-X holdings after hearing Walsh's shocking announcement.

Mutual funds owned the bulk of Bre-X stock, but they didn't have the most to lose; their holdings were large and relatively diversified. Some thirteen thousand individual investors who held Bre-X in their own individual accounts — especially those who had bought Bre-X on margin (credit) — were bound to be wiped out once trading resumed. Selling would be fast and furious. The brokerages would look after their own interests by covering as many margin accounts as possible. Investors still holding the stock would have very little time to react. But no one anticipated they would have just fifteen minutes.

A quarter of an hour. That's all it took to knock eighty per cent off Bre-X's market value. When the Toronto Stock Exchange lifted its trading embargo on 27 March, the stock nosedived to $2.50, flushing $3 billion down the drain. Then the TSE's computer system overheated and crashed. Exchange officials explained that their hardware was twenty years old and couldn't handle the heavy volume.

Greg Chorny still had 350,000 Bre-X shares sitting in his account. Although the Ontario lawyer had already made about $35 million selling Bre-X stock, he was livid with the TSE for allowing trading to open that day. "The TSE wasn't worried about individual Bre-X shareholders," he says. "It was worried about the members sitting on its board. There were rumours that brokerages like Nesbitt Burns were actually imperilled because of their exposure to Bre-X. There was a ton of Bre-X stock out there that had been bought on margin. Some brokers had given their clients full credit with no collateral. They couldn't demand payment if the stock was cease-traded, so they demanded the exchange get it moving."

Others argue the TSE really had no choice but to resume trading, as soon as possible. Speculation was so great, exchange officials said, that an off-counter "grey market" had developed in the United States. In order to keep the trading accessible to everyone, the exchange felt compelled to put Bre-X stock back into play. But it's doubtful that many retail investors had time to dump their stock and cover their margins.

Bre-X was headed straight for the scrap pile, dragging the rest of the junior mining sector along with it. Every small exploration outfit saw its stock drop precipitously following Walsh's big announcement. Companies with Indonesian properties were hit especially hard, many losing half their value. The Vancouver Stock Exchange and the Alberta Stock Exchange, both dominated by junior resource companies, suffered their largest daily declines since the global stock market crash of October 1987. Every investor headed into the Easter long weekend, shocked and dazed, wondering what would come next, praying that the debacle was over. It should have been, but it wasn't.

Busang refused to die. Many observers — engineers, miners, executives, journalists — couldn't accept that the deposit had been faked. The deception endured as the world waited for Strathcona's final, definitive statement in early May. Strathcona was at the site, drilling more holes, preparing to evaluate the fresh core. Meanwhile, back in Canada, there was plenty of finger pointing, innuendo, denial and conspiracy talk to keep people occupied and confused.

Bill Stanley had given up counselling his fellow thread-watchers to stick with the facts and not get carried away with weird ideas that the Busang deposit was real. He'd already been flamed for his efforts, and now an anti-Bigdude cabal was patrolling the thread, posting boneheaded bounties on Bigdude's head. "Dud, the gig is up," wrote one poster. "I am personally going to make sure you are taken down. The gold is there, you know it, and you are a punk that has helped a lot of people lose money." The threats worked. Afraid of being outed, Stanley slowed Bigdude's output, taking care not to bad-mouth Bre-X.

Privately, he fumed. The paranoid Internet chatter had him twisted in knots. "Everyone is missing the point," he said. "Bre-X never twinned its holes. No one had any independent evaluation of Busang until Freeport. The skeleton core I brought back from Busang tested negative. Then there's de Guzman's disappearance. I'm just totally suspicious of Bre-X right now."

The class-action lawyers descended; seven angry suits had been filed against Bre-X, on behalf of shareholders in New York, San Francisco, Seattle, Atlanta, Calgary, Montreal and Toronto. The allegations were set out in loose, catch-all terms and directed at a short list of Bre-X directors. Other defendants were quickly added, deep pockets such as Kilborn, the regulators, and the brokerages that had touted Bre-X stock.

None of them offered Bre-X any support. Damage control demanded they distance themselves from the debacle, and their statements did not help Bre-X's credibility. Kilborn's Canadian parent, SNC Lavalin, finally set the record straight on its role at Busang. Kilborn was "very downstream" of the core sampling process, said

Robert Racine, an SNC Lavalin spokesman. It "did not drill, did not take the samples nor did it assay those samples."

The TSE had already defended its decision to re-open trading. Now it was taking flak for not properly screening Bre-X and giving it blue-chip treatment instead. Groups such as the Pension Investment Association of Canada, which represents $365 billion in pensioners' money, asked why Bre-X had landed on the exchange's composite index of its three hundred top companies. The TSE's lame response — that Bre-X qualified because it possessed a high market capitalization, because it had money, nothing else — confirmed there were flaws in the regulatory system. (Bre-X ignored TSE guidelines regarding its board of directors. The exchange very sensibly asks that listed companies make sure that senior officials do not dominate their boards. "Un-related" directors are supposed to form the majority; this wasn't the case with Bre-X. A helpful TSE official pointed out that only two-thirds of companies on the 300-index actually adhere to that guideline.)

The analysts became uncharacteristically silent. Bianchini did not return telephone calls. He couldn't talk, as he was being sued. Chad Williams was likewise unavailable for comment; there would be no explanation for how he had concluded that activities at Busang were "systematically executed in a way that can be easily verified." Michael Fowler, the bullish analyst from Levesque Beaubien, also said he couldn't talk. Those analysts who did express belief that Busang was a scam either demanded anonymity or were weighing in for the first time.

Fund managers stuck with $3–Bre-X stock tried to sound positive. "This is a buying opportunity," Vic Flores, an American gold fund manager, said in early April. Freeport had been too hasty releasing results from its due diligence campaign. Never mind the TSE and the U.S. Securities and Exchange Commission required Freeport to make the statement. "How could they damn the deposit with so little work?" Flores asked. "The due diligence is something that will take maybe two months, not two days. To say after a handful of drill holes [that] there's no gold is irresponsible." Two per cent of Flores's gold fund consisted of Bre-X stock. He was "hoping for a bounce" in the price in the future. "I was firmly in the camp of believers," he told me. "I thought that Freeport made a mistake."

The idea that Freeport, not Bre-X, was wrong, developed into a full-blown conspiracy theory. It got legs when newsletter publisher Michael Schaefer passed story along to subscribers. "A couple of weeks ago," he wrote, "I was told that Freeport and certain Indonesians were going to initiate a campaign to totally discredit Bre-X and the Busang Gold Deposit. I was told that Freeport would downgrade the Busang Deposit and then take over the company at a $3.00 to $4.00 share price."

Schaefer "can't remember" who gave him the information, just that it was some time in February, as Freeport began drilling at Busang. He was still speaking regularly with David Walsh at that time. Coincidentally, on 22 February, Walsh told Canadian Press reporter Valerie Lawton that "I firmly believe there are sources involved who are either shorting our stock or are considering Bre-X as a takeover target."

Schaefer says the story first seemed absurd, until Bre-X's integrity over Busang came into question and its stock fell. His bulletin was posted on the Bre-X discussion thread within days of publication and spread like wildfire. Pro–Bre-X posters latched on to the delicious idea and promoted it tirelessly. Freeport "DELIBERATELY chose to drill and sample in such a way as to present smaller samples," wrote a poster called R.M.C. "It seems to me that Freeport (now having 15% and being operator) wanted more. The corporate culture of Freeport indicates that this is not out of the realm of possibility."

Mikesloan, Bre-X's net weasel, claimed that "a huge clandestine operation" was "being very carefully orchestrated behind the scenes. . . . This is an attempt to take over the company." As the Strathcona deadline approached, and Mikesloan grew more strident, his list of conspirators grew to include Barrick, Canadian regulators and most of the country's business press.

Ironically, the daily media began mining the same Bre-X thread for information. Reporters from *The Globe and Mail, The Financial Post*, and the Canadian Broadcasting Corporation began e-mailing regular posters, asking for interviews. In some cases, newspapers printed rumours picked from the thread. *The Financial Post*, for example, repeated anonymous chatter deriding Freeport's assaying methods. Canadian newspapers and radio stations also repeated chat group gossip hinting that that Jim Bob Moffett had been forced to resign his post as Freeport president. Approximately seventeen million Bre-X shares traded hands the day the rumour was posted, driving the stock price from $2.40 to $5.75. One gold analyst contacted by reporters said that if the rumour was true, "then it's basically an admission that there's been a huge problem, that Freeport made a mistake and Bre-X is vindicated." Although Freeport quickly denied the story, Bre-X stock continued to trade at unjustifiably high prices.

What should have been a dead stock continued to attract massive interest from investors. Some of the trading in Bre-X was done in small blocks through Canadian discount brokerages. Lots of Canadians were willing to put a few hundred dollars on Bre-X; the stock was cheap, and the upside — assuming Busang was real — was enormous. For many, it was a bit of harmless fun, like betting on a Super Bowl long shot.

But there were large buy orders coming from Europe, Hong Kong and New York. According to some observers, this indicated that institutional investors were covering their short positions and reaping huge profits. Short-sellers make money by arranging to borrow stock that they believe will fall. If the stock drops, they are entitled to pay for the stock at its new level, sell it back to a brokerage at the original price, and keep the difference. If it climbs, of course, the short-seller loses.

It's a gamble, but short-selling is legal and very common. And Bre-X was a short-seller's dream, because of Busang's dubious nature. Rumours that Freeport had found nothing at Busang leaked from Jakarta five days before the company officially released its negative drilling results. This gave sharp players time to take short positions on Bre-X stock, which was was still trading at around $15. On 26 March, the stock plunged eighty per cent, and the short-sellers began covering. Some made incredible profits. The New York-based brokerage Oppenheimer & Co., for example, claimed to have made $100 million short-selling Bre-X.

In a last-ditch effort to keep his company afloat, Walsh went into promotional overdrive. There was no chance that Busang was a fake, he said, again and again. "I am 120 per cent confident that the gold is there and that there has been a colossal screw-up," Walsh told *The Toronto Star*. Busang could not have been salted, he insisted. "It's a physical impossibility. You'd have to have an army of people." He blamed the meltdown on a tangle of anonymous bogeymen; greedy short-sellers, take-over artists, basically anyone not sympathetic to his cause. A "well-oiled disinformation campaign" was at work, he explained, and it was aimed at dragging down Bre-X. In fact, the record shows that Walsh had embarked on a disingenuous propaganda campaign of his own making.

First, Walsh cast doubt on Strathcona's "preliminary" conclusion, which found that the size of the Busang deposit had likely been overstated due to invalid samples. Strathcona's statement was, Walsh said, "to the best of my knowledge, based only on Freeport data and information from Freeport." This was simply not the case. Strathcona had reviewed Kilborn's intermediate feasibility study, which included the troubling metallurgical reports, and it looked at other information gathered in Toronto and Jakarta. Strathcona clearly mentioned this in a letter to Bre-X dated 26 March. Later, the company released a document reiterating that its conclusion was based on Freeport's drilling results *plus* "our own observations in reviewing the reports and information that we had received. . . . We noted a number of observations that we considered to be indicative of potential problems with the Bre-X sample data."

Walsh neglected to mention another salient point in his drive to keep Bre-X alive. Attempting to quash mounting speculation that drill core from Busang had been salted on the way to the assay lab in Balikpapan, Bre-X released a flow chart of its handling procedures.

Bre-X falsely indicated that the core travelled from Busang directly to Balikpapan, without stopping. There was no mention of the large Samarinda warehouse — the salting shack — where Busang core sat under lock and key for days, even weeks, before actually moving along to the lab. In David Walsh's mind, the warehouse never existed.

Reality finally caught up with him on 4 May, when a group of Strathcona representatives arrived in Calgary with their final report. The cover letter said it all:

CONFIDENTIAL

Dear Mr. Walsh,
Enclosed is our report covering that portion of our technical audit of work on the Busang property. We very much regret having to express the firm opinion that an economic gold deposit has not been identified in the Southeast Zone of the Busang property, and is unlikely to be. . . . The magnitude of the tampering with core samples that we believe has occurred and resulting falsification of assay values at Busang, is of a scale and over a period of time and with a precision, that, to our knowledge, is without precedent in the history of mining anywhere in the world.

14

FALLING

No one sitting here tonight can agree on what the word Busang really means. There are at least three definitions. To me, Busang is a wild cat, an elusive, mischievous creature that steals our chickens. When our people went looking for gold along the rivers, they would occasionally glimpse the Busang, before it darted into the jungle. Now it is seen more frequently, and it is bold.
– Ngang Bilang, Dayak elder, May 1997

PIERRE TURGEON TOOK HIS OWN LIFE on a cold Thanksgiving weekend. A veteran stockbroker at Nesbitt Burns, Turgeon had talked about killing himself for months following the Bre-X collapse. His clients were unhappy, their money was lost, and Nesbitt Burns refused any responsibility. *Sorry, that's your problem, caveat emptor.* Turgeon's personal situation was precarious. He couldn't bring himself to tell his elderly parents that he had blown it, that the meltdown had left him with nothing, that his career was in ruins. So he jumped. He stepped off the balcony of his seventh-floor apartment in downtown Montreal and he was gone.

His death passed without comment. There was no obituary, no public funeral, no articles in the press. Turgeon was forty-six years old and left no wife or children behind. No one at Nesbitt Burns will discuss the tragedy. Nesbitt Burns will no longer discuss anything associated with Bre-X, including employees who killed themselves over it.

George Diekmeyer will. Diekmeyer was Turgeon's friend, and a client. The sixty-four-year-old man with deep-set eyes and salt-and-pepper hair lost $300,000 when Bre-X crashed. The money represented most of his life savings. For a time, he also thought about killing himself. "One is attached to life, but only to some extent," he says, his voice cracking slightly. "What was it Nietzsche said? 'The thought of suicide gets one through many a bad night'?" Six months before the crash, at Turgeon's suggestion, Diekmeyer had put his investment portfolio on the line and bought Bre-X stock. He held onto it despite mounting evidence that the company was in trouble. Now he sits in his house in suburban Montreal, wondering how to meet his mortgage obligations and pay his property taxes, worrying that he and his wife, Gerda, may have to liquidate everything they worked so hard to acquire.

The Diekmeyers have "severe problems" financially. They are both in their mid-sixties and have little time to play catch-up. Gerda works as a physiotherapist at a local hospital. Her income covers their day-to-day requirements such as food, electricity and clothing. George doesn't blame Turgeon for the initial idea; he knows that his friend was only recommending what had come highly touted from Bianchini. But he does blame him for not being alert to the developing scandal.

"Pierre and I spoke repeatedly about this," says Diekmeyer. "He kept asking me how we could have been so blind. But we both knew the score. I know I would never have bought stock in Bre-X had the chief gold analyst at Nesbitt not recommended it. I was persuaded that Egizio Bianchini knew things about Bre-X and its property that no one else had access to. I read his reports. They were very powerful, very positive." Turgeon, he thinks, "was torn apart by a terrible conflict. He wanted to talk about what had really happened. But he had a duty to Nesbitt Burns which barred him from talking to his clients about Bre-X. This put him in a tremendous bind, and he couldn't see how to go on. Can you see how confused he was? He told me several times that he was considering jumping from his balcony. I urged him not to. I suggested he seek professional help. I tried appealing to his intelligence, suggesting that he walk out of the industry, close the door and move on. But he just couldn't do that."

Diekmeyer has found life difficult too. He was deeply depressed after Bre-X imploded, and struggled to retain his self-respect. "You fight for space, you fight for a better place to live, you fight for this and that," he says. "And sometimes you wonder if it's all worth fighting for." He's feeling better now, months after the meltdown, but his anger hasn't diminished. "I know people look at me and think, 'So, this is the man who got wiped out. How could he be so stupid? How could he have gambled everything?' And I don't know what to tell them, except that I didn't think I was gambling."

The sad irony is that Diekmeyer has exercised caution most of his life. He's not some wild speculator, or a naive yokel new to the stock market. He used to be a broker once, with Burns Fry Ltd., a predecessor of Nesbitt Burns. He's been a wary investor, spending wisely, learning from his mistakes, choosing his investments carefully. "My goal has always been to avoid risk, to look for a sure thing," he says. "My gut instinct has always tended towards security."

It's an impulse he's had since childhood. Diekmeyer grew up in northern Germany during the Second World War. He remembers hearing the American bombers fly overhead at night, on their way to set Munich ablaze. "The sky would light up, as if there were fireworks going off, only you knew that it was no show, that a city was being destroyed," he recalls. "The sight of all those planes, the sounds of the bombs exploding, it was just unforgettable."

When the war ended, he couldn't wait to get out of Germany. All the Nazi propaganda of the 1930s and 1940s and the Aryan myths he had grown up with left him feeling disoriented and insecure. In May 1955, Diekmeyer bought a $200 steerage-class ticket to a new life. He boarded a thirteen thousand tonne ship in Bremen and set sail for Canada. Two weeks later, his ship landed at Quebec City, and Diekmeyer, alone and with little English to guide him, continued by train to Toronto. He'd heard that the city was as good a place as any for a young immigrant to get a fresh start.

Toronto seemed "run-down, disorganized, pretty dreadful," Diekmeyer recalls. He was startled to find a large lower class scrabbling about the city, trying to make ends meet. "Tramps slept in subway stations," he says. "I was shocked at this. I thought things were supposed to be better over here." He ended up in a dilapidated $5-a-week rooming house at the corner of Bathurst and College streets, a flickering light bulb dangling from the ceiling of his tiny room. He lost half his savings after naively lending $25 to a stranger he'd met one night in a bar. The man promised Diekmeyer he'd pay him back the following week; naturally, he never saw him again.

Diekmeyer latched onto a job stacking department store catalogues, but was laid off on Christmas Eve. After a few months of living hand-to-mouth, he ventured east to Montreal, where he found a job with Canadian Pacific Railways. It was dull work, entering numbers in ledger books, but it paid the bills. He stayed for three years, eventually earning a commerce degree at night school. He married Gerda and bought a small house for $7,000. They began planning for children and eventually had three. They felt safe.

George first took an interest in the stock market in 1958, while working at the CPR. He heard about a gold mine down in Cuba, took a flier on it, and lost. But the experience didn't sour him. He continued to make small investments in the oil and gas sector. "It wasn't terribly rewarding, but I stuck with it," he says. As Diekmeyer's job situation improved, so did his stock portfolio. He began to buy into large, blue-chip companies and sold options on his shares. Option trading allows a stockholder to sell someone the right to buy his shares, at a certain price, on a specific date. The option buyer may either "exercise" the option at the agreed-upon price, or allow the offer to lapse. Either way, the seller makes money.

The system helped Diekmeyer purchase a four-bedroom house

in suburban Beaconsfield. Soon he was devoting so much time to the market that he decided to become a broker. In 1980, after completing the Canadian Securities Course, he joined the investment house Burns Fry, which later merged with Nesbitt Thomson Inc. to become Nesbitt Burns. Diekmeyer spent five years there, never really fitting in. "My goal was to have as few clients as possible, be very familiar with them, and give them responsible advice and do a good job," he says. "That to me was the whole reason for being in the business." Instead of trying to sell them on every new stock offering that came down the pipeline, Diekmeyer urged his customers to follow his own conservative strategy. As time passed, he realized that this approach no longer fit the Burns Fry agenda.

"It seemed to me that a tremendous change took place," Diekmeyer says. "A lot of cynicism starting coming into play. There were still people who took the business very seriously, not necessarily geniuses, but they would be very conscientious about what they did. But then an entirely new philosophy set in with the arrival in large numbers of the MBAs. They started actively managing people, [giving them] standards to meet. That is, actively encouraging them to add twenty per cent each year to their turnover. I thought it was just horrible. Because it completely perverted the idea that you were there to serve your clients. You were serving your bosses, the guys who were raking in the twenty per cent increases that you were making."

Diekmeyer left Burns Fry in 1985 and turned his account over to Turgeon to manage. The two men were from completely different backgrounds, and of a different generation, but they shared a similar investment philosophy. Turgeon was depending on a limited number of clients and did not seem concerned about making quick money. "Pierre was trying to do a good job," says Diekmeyer. "I appreciated his diligence and honesty." He thought they made a good team.

At some point, Diekmeyer became convinced that most stocks were hopelessly overvalued, the result of market hype. As a result, he began to invest only in gold-mining stocks. Gold, he reasoned, was permanent. "The price fluctuates, but its real nature never changes," he says. "It is essentially always the same. So I looked for well-managed mining companies which were actively trying to increase their reserves."

Eventually, he sold all his other stock and invested exclusively in Placer Dome. He held about thirty thousand shares and started selling options. The strategy seemed foolproof. It was, he says, his "atomic bomb shelter." Even when Placer's stock fluctuated, he continued to make between $40,000 and $80,000 a year just by selling options. He seemed to have it made. His days were spent swimming, jogging and talking to Turgeon.

They watched Bre-X from a distance. "It wasn't my game," says Diekmeyer. "Things were moving so fast, it was hard to follow what was really happening. It's like when you're driving along in a car, and you see something flash by, and there's a crowd of people running behind it. Do you turn around and chase after them? I wasn't that interested."

That changed in the fall of 1996, when rumours first surfaced that Placer was trying to close a deal with Bre-X. The story was that the company was going to offer one Placer share for every Bre-X share, plus a small cash premium. This got Diekmeyer thinking. He could buy Bre-X shares on margin, using his Placer stock as collateral. If Bre-X was taken over by the Vancouver-based company, he'd end up with more Placer shares *and* scoop the premium the company was offering. The only risk he could see was that Placer might be outbid by a rival company. If this happened, he would still end up with shares in another major mining company, or so he imagined.

At Turgeon's suggestion, Diekmeyer started nibbling at Bre-X. He started slowly, buying three thousand Bre-X shares for two RRSP accounts belonging to him and his wife. When Placer Dome made its formal offer to buy Bre-X, he dived in head first, acquiring, on margin, another ten thousand Bre-X shares priced between $20 and $22. In February, after Freeport announced its deal with Bre-X, Diekmeyer bought another two thousand shares, this time at $24.

Placer had lost the battle for Busang, but Bianchini was still recommending Bre-X stock. Diekmeyer didn't worry when Felderhof bragged that Busang hosted an incredible 200 million ounces. "I thought that that was just hype," he says.

Then de Guzman "fell." Rumours spread that Freeport hadn't found any gold at the site. Diekmeyer began to lose confidence. "I talked to Pierre. I said things don't look good. The stock had dropped to $17 and I was becoming quite apprehensive. Pierre said, 'Who cares? Geologists are replaceable. He had hepatitis.'" As for the Freeport rumours, Turgeon dismissed them out of hand, reading from Bianchini's infamous "The gold is there!" recommendation. "That report persuaded me not to sell," says Diekmeyer. "I didn't think Bianchini could make that statement unless he knew it was true."

He was wrong. Bianchini apparently didn't have a clue what he was talking about. One day after Bianchini's report, Strathcona reported that the size of the deposit was overstated, and Freeport confirmed it had found "insignificant amounts of gold" at Busang. Bre-X stock melted and investors such as Diekmeyer were left stranded. Nesbitt Burns wanted Bre-X off its books. The brokerage quickly enforced its margin calls, reaching into its clients' accounts. Diekmeyer watched helplessly as the brokerage seized his eight thousand Placer shares.

"I couldn't believe what was happening," he says. "It was unconscionable. [Nesbitt Burns] recommended the stock to almost double in price. The next day, I lost everything, without any chance of recovery. I asked Nesbitt to at least give me an opinion on Bre-X. And they refused, which I think is a total cop-out and absolutely sick. They were only concerned with covering their own asses." Diekmeyer points out that had Strathcona came back later and reported that Busang was genuine, he could not have resumed trading. His assets were gone. His brokerage had sold him out.

Diekmeyer was furious. "I called Nesbitt and I said look, leave my account alone, let's talk about how to minimize the damage," he says. "I suggested that we find some kind of settlement which would allow me to continue in the business, because without my capital, I can't operate. It's like when a pilot loses his sight, or a baseball player loses a leg. It's not very good for their career. But Nesbitt refused to discuss it."

Nesbitt's exposure to Bre-X was so great, some say, that its viability was threatened. Hence it made thousands of margin calls without waiting for the final, definitive drilling results from Strathcona. Others argue that since everyone else on Bay Street was writing down Bre-X's value, Nesbitt Burns was just following the trend. Whatever the case, Nesbitt Burns blew its Bre-X coverage. It got the facts wrong, it suggested that Busang was a sure bet, it resisted any notion that something was wrong with the deposit, and then it bailed out.

The brokerage denies any responsibility for the losses sustained by its clients. Nesbitt Burns faces at least two multibillion-dollar class-action lawsuits. These claim that the firm acted negligently by recommending Bre-X stock and thus breached its fiduciary duty to its clients. These are difficult allegations to prove. Brokerages issue fine-print warnings on all their stock reports, full of caveats designed to limit their responsibility in case a "Buy" recommendation goes bad. In Nesbitt Burns's case, the wording was that it "makes no representation or warranty, express or implied, in respect thereof, takes no responsibility for any errors and omissions which may be contained herein and accepts no liability whatsoever for any loss arising from any use or reliance of this report or its contents."

While claims for breach of fiduciary duty have often failed in the past — investors are, after all, independent and able to make individual decisions — cases of negligence have been successfully argued. There is always risk when investing in securities. But brokerages have a legal responsibility to define that risk. Did Nesbitt Burns do an adequate job?

The brokerage stands by its work. After a brief respite, Bianchini was back at work, visiting gold properties in the United States. His reputation as an analyst has been damaged, perhaps permanently. Even so, Nesbitt Burns insists he acted professionally at all times.

There was no negligence. Paul Gammal, a Nesbitt spokesman, told the *Washington Post* that "our analysts rely on publicly available information that is released by reputable companies, but in this case there is fraud. Our experts are not experts in uncovering fraud."

Admitting even the slightest responsibility for its clients' losses isn't in the cards. Nesbitt Burns takes the position that it, too, was "a victim" of Busang. That doesn't help Barbara Horn, the Calgary based Nesbitt Burns broker who relied on Bianchini's "expert" advice. "I don't understand geology," she says. "I was depending on our research department. And Egizio, it really bothers me . . . he's a geologist, and he went down there to Busang, and he came back with all his great stories. Whenever there was any doubt about the deposit, I kept saying to people that the gold isn't going to get up and walk away. But it did. So I've lost clients. People lose confidence in you."

While Horn's personal exposure to Bre-X was small, other Nesbitt employees weren't so lucky. Rumours persist that approximately forty Nesbitt employees were wiped out after getting stuck with worthless Bre-X stock. Michel Mendenhall, a Nesbitt broker from Ottawa, was a major Bre-X buyer. He knew David Walsh from a bad investment he'd made with him years earlier and had a direct line to his office, which he used frequently. Mendenhall ended up with more than two million shares. To finance his spree, he borrowed from Nesbitt. Then he followed Bianchini's advice and rode Bre-X all the way to the end. When the company imploded, he was left with a $3.7-million shortfall. Mendenhall was forced to seek court protection from his creditors.

Back in Montreal, Diekmeyer continues to speak quietly about the past twelve months, what a struggle it's been. "Losing all that money hurt," he says. "I know it's probably too late for me to ever recover that. But I think my biggest problem is with the inhumanity of the people in that organization. Nesbitt never considered what impact its actions might have on people. And they were huge. Look at Pierre. Horrible, horrible consequences. But their people want to say, 'Oh well, you lost, that's bad luck, that's the end of the story.' If it is the end, then God help us, we're in for a really bad experience.

Because luck shouldn't have anything to do with it. We are talking about investments, not games of chance."

One by one, the reminders fall. The Three Greenhorns is long gone. David Walsh's old watering hole has been replaced with a nondescript Italian restaurant. The Bre-X building sits empty; the large, golden letters that once spelled out the company's name have been removed. Walsh doesn't visit Calgary much. There's little need; Bre-X was forced into bankruptcy in November 1997, after Walsh and his colleagues emptied the company's bank account, spending $1 million a month on legal fees, salaries and "performance" bonuses. Bresea, which remains solvent with approximately $26 million in cash, was placed under the control of a court-appointed receiver. Rebuffed in his attempt to turn Bresea into an oil and gas outfit, Walsh retreated to Nassau, where he and Jeannette minded a $10-million renovation to their beach-front spread. When he is in Calgary, Walsh moves from bar to bar, keeping a low profile, the Salman Rushdie of mining. As his old buddy Fast Eddie Schiller used to say, you can't hit a moving target.

John Felderhof's name has been erased from the mining industry's pantheon. He surrendered his Prospector of the Year award a few weeks after resigning from Bre-X in May 1997. Unlike Walsh, Felderhof hasn't budged from his own Caribbean exile. He won't leave the Cayman Islands. His assets — cash, houses, cars and a hulking Sea Ray speedboat — have been frozen by local authorities, pending the outcome of a Canadian class-action suit.

Walsh and Felderhof don't speak to each other any more. The two men have split ranks. Walsh readily admits that a fraud took place, but insists that the venue was distant Kalimantan, not Canada. It didn't happen under his nose. Forensic Investigative Associates Inc. (FIA), the Toronto-based sleuths Bre-X hired to assign blame, concluded that Felderhof's role was "an open question." They did, however, single out geologists Michael de Guzman, Cesar Puspos, Manny Puspos, Jerry Alo and Bobby Ramirez. The five Filipinos began salting core samples pulled from Busang in December 1993 and continued their covert campaign until March 1997. No one else knew.

FIA's 434-page report was completed in October 1997 and was finally released to the public by Bre-X's Trustee in bankruptcy early in 1998. It is a deeply flawed piece of work, relying on rumours and gossip from questionable sources. Indeed, FIA admitted that "not all information [contained in its report] is sufficiently credible, cogent, or complete to permit reasonable inferences and conclusions to be drawn from it." Nevertheless, its judgments are sweeping.

The report began with the notion that neither Walsh nor his colleagues in Calgary knew that their prized gold deposit was salted. But since FIA's inquiry was limited to Bre-X's activities in Indonesia, the argument lacks substance. FIA reported that "the head office in Calgary did not have a representative of the Indonesian technical group located there and so there was very little reason for any information relating to the operational decisions to be transmitted to Canada." It's hard to imagine that Walsh, the president and chief execitve officer of a company claiming to have found the largest gold deposit on the planet, did not take a profound interest in what was happening at the site.

But assume FIA was correct. It's clear that Walsh paid no attention to his Indonesian operation. He completely neglected his duties, not to mention the best interests of his cherished shareholders. "The point isn't simply whether Walsh knew what was going on at Busang," notes disgruntled investor Greg Chorny. "It's that he should have known. He paid for all those metallurgical reports that mentioned the curious shapes and sizes in the gold matter. He paid for a technical report that suggested the whole assay process was flawed. There was all kinds of information that he sat on."

To help absolve Walsh of any responsibility for the fraud, FIA pointed to testimony from a young Canadian geologist who worked briefly at Busang between 1996 and 1997. Steven Hughes, a recent graduate of Saint Mary's University in Nova Scotia, allegedly told FIA that he did not believe Walsh "knew anything about the salting."

No reason was given for this opinion. According to FIA, Hughes believed "that both de Guzman and Cesar Puspos were involved in the salting and that it was a conspiracy among the Filipino geologists and metallurgists." Four more Canadian geologists who spent time at either Busang or Samarinda pointed the finger at the Filipinos, with nary an incriminating word about the Calgary connection. One of those other geologists was a friend of Hughes', named Trevor Cavicchi.

I first heard of Hughes and Cavicchi in early May 1997. Out of the blue, Hughes called me at my home. He told me he had been employed at Busang for a time and briefly described the working conditions there. Then he offered to sell me his "complete" story. "How much are you willing to pay?" he asked. I told him I would have to think about it. A few days later, I spoke with Cavicchi. He was also attempting to flog his story. He told me there were other people interested in obtaining "exclusive" access to his memory bank, and that if I wanted his services, I had better act quickly. Several weeks later, Cavicchi informed me that they had received a "substantial" offer from another source, and would not speak to me. (FIA interviewed Hughes on 26 May and 17 June, 1997, and interviewed Cavicchi on 1 May, 1997. FIA did not mention whether its sources were compensated.)

FIA promoted a second premise, that de Guzman jumped from a helicopter on 19 March, 1997. "We believe he killed himself when faced with the prospect of having to be exposed for salting Bre-X samples," FIA noted. "All the evidence that we have reviewed points to [this] conclusion." Unfortunately, the only "evidence" contained within the FIA report is circumstantial and anecdotal. FIA investigators chose to emphasize a story passed along by Rudy Vega, who accompanied de Guzman from Balikpapan to Samarinda on 19 March:

While en route to the airport, Michael made the following statement at least four times: "Shit, I should not have done this. This is the second time." When I asked him what had happened,

Michael said he had fallen asleep in the bathtub after drinking an entire bottle of cough medicine. It was my impression that Michael had tried to commit suicide that night by drinking the bottle of cough medicine in hopes of falling asleep in the bathtub and drowning.

FIA claimed that de Guzman had displayed "anguished emotions" on the morning of 19 March. This assumption is based entirely on Vega's strange story and plainly contradicts other testimony contained within the report. What's more, Vega's tale is the foundation of what FIA labelled as "evidence capable of demonstrating consciousness of guilt." De Guzman, a liar of pathological proportions, was instantly overcome with regret and ended his life. Or so FIA wants us to believe. It's nonsense.

There is another complication. "We are informed that there was difficulty in Manila regarding the use of fingerprints to confirm the identity of [de Guzman's] cadaver," FIA noted, in its report. Yet FIA was "satisfied on the basis of the information we have reviewed that the cadaver recovered from the jungle is that of Michael de Guzman." Why? Because that's the story that fits. It's convenient. And that, it seems, is the whole premise behind David Walsh's pathetic defence.

John Felderhof's story has more turns than the Mahakam River. He may have been the crucial link between Walsh and de Guzman, but he was rarely at Busang, his lawyers insist. He didn't know that bags of core samples in Samarinda were systematically being opened. He knew nothing about any tampering; results from a lie detector test were offered as proof. But Felderhof didn't stop there; he actually promoted the idea that the hoax was a hoax.

"I continue to believe that there is a significant amount of gold at Busang," Felderhof wrote in a statement delivered to the press in July 1997. Michael de Guzman couldn't have salted Busang, he insisted. "I still find it very hard to believe that as a respected professional geologist, he tampered with Bre-X's gold samples." The "so-called 'red flags' that allegedly should have alerted me...to the effects of tampering" didn't exist, he added. Busang was a geological treasure, akin to other large tropical deposits, showing "high percentages of free gold, a common feature." There were just "a few anomalies" in the size of the grade, he said. Most of the assay results "typically showed 1-3 grams of gold per tonne. This too was consistent with my experience." What's more, the whole-core versus split-core debate was nonsense. "It was a very simple procedure for Freeport, Strathcona, or anyone else to go back and redrill 'twin holes' to double check Bre-X's claims." Felderhof went on to say that "many other companies had done extensive work on the site." Not true. Bre-X wouldn't allow it. Until Freeport forced its way onto the property, no one was ever allowed to drill on Busang.

Months later, the story began to change. "John is beginning to come around to the idea that Busang was salted," his Canadian lawyer, Joe Groia, told me in January 1998. Felderhof's earlier prognostications regarding the deposit's size were never meant to be taken *seriously*. They were "amorphous statements of optimism," according to a defence motion filed on behalf of Felderhof in an American district court. "Reasonable investors should not consider 'soft' opinions or vague, optimistic predictions important or relevant in making investment decisions."

Good advice. Except Felderhof was anything but vague when making his claims. "I feel very comfortable with a potential 200 million ounces," he had said. "I'm 110% confident the gold is there." Now he's not so sure. Why believe anything he has to say?

The class-action pitbulls have plenty of gristle to chew on. Throw in the rest of the defendants — analysts, engineering firms and the regulatory bodies that allowed Bre-X to trade — and chances are that some of their allegations will stick. Shareholders may get a few dollars back, unless, of course, the lawyers crack up. Infighting threatens to sabotage the largest class-action suits directed against Bre-X. A fifteen-firm mega-suit filed in the litigation haven of Texarkana, Texas, appears headed for nowhere. Early on, two of the mega-suit's lead attorneys, Paul Yetter and Tom Ajamie, were forced to resign from the high-profile Houston-based law firm of Baker and Botts after refusing to dump the Bre-X case. The suit was deemed "too controversial" for a pin-stripe firm such as Baker and Botts, says one insider. "The corporate lawyers thought it was unseemly. The fact was, two of the investment houses named in the Bre-X suit were also clients of the firm. [Yetter and Ajamie] were told to walk away from the suit immediately, or leave the firm."

Problem solved; however, a rift developed among some of the American attorneys preparing the mega-suit. "We're supposed to be working together on this for the sake of our clients," says one disgruntled mega-team member. "Some of the guys seem more interested in grandstanding in front of clients. So, yeah, they're keeping secrets from each other. It's stupid, but I guess that's what happens when you get this many lawyers working together on a common cause." Ajamie, who spent close to three months in Indonesia while working on the case, admits that there's been some squabbling. "This is going to be a long, hard case," he says. "We're going to get bogged down by endless delay tactics from the defendants, so there's no point in creating more problems for each other."

The class-action suits were filed immediately after the meltdown — in some cases, even before the fraud was announced — but it could take years before anyone associated with Bre-X actually ends up in court. In the meantime, unsuspecting investors will continue to be deceived by unscrupulous mining promoters. At least two other "misrepresentations" have already been documented, fast on the heels of Busang. Investors remain gullible. But the rip-offs also underline the need for tighter regulatory requirements and more accountability from stock analysts. Unfortunately, little has changed to prevent more bogus mining promotions from being traded in the future.

In May 1997, Vancouver-based Delgratia Mining Corp. was forced to admit that samples from its so-called Josh deposit in Nevada were salted with gold. The company had announced that the site contained at least five million ounces of gold, and Delgratia stock, which traded on the NASDAQ, shot to US$34.75.

Nevada's state mining officials wisely questioned Delgratia's claims, calling them "premature," and the company's shares lost half their value overnight. Delgratia stood by its findings for weeks, noting that two independent labs had confirmed a large presence of gold in its assay samples. But a third study, commissioned at the state's insistence, revealed that there was "insignificant gold contained in the Josh deposit and that any gold detected beyond background amounts was introduced into the samples after they had been collected at the drill."

Tearing a page from David Walsh's book of excuses, Delgratia president Charles Ager denied responsibility for the fraud and vowed to help bring the real perpetrators to justice. Angry shareholders are still waiting. By year's end, Ager and the rest of the Delgratia board had been replaced by a new group of executives, and the Josh property was written off at a cost of $22 million, leaving the company with less than $9 million in its bank account. Eight class-action suits have been filed in the United States and Canada, alleging that Delgratia officials misled investors. Delgratia's new board plans to "vigorously defend" the charges; unfortunately, this won't alter the fact that a fraud was committed, causing $275 million in market capital to vanish.

Then there was Golden Rule Resources Ltd., a Calgary-based company listed on the TSE. Like every other junior mining outfit in Canada, Golden Rule exploited the speculative bubble, quickly raising large amounts of cash in the marketplace, even in the absence of any drilling results. In late 1996, a mysterious "whisper campaign" touting one of the company's far-flung properties in Ghana worked its way inside Canadian brokerages and mutual fund offices. The story was that Golden Rule had found millions of ounces of gold at its site.

This was absurd, since the company had only completed some initial surface work. Early results released by Golden Rule allegedly showed a stunning 8.8 grams of gold per tonne along a 100-metre-long trench dug along the surface of its Ghana property, called Stenpad. This was promising, but not enough to build a resource estimate. Nevertheless, big numbers began flying about in the press. During a January 1997 exchange with a *Globe and Mail* investment reporter, Shirley Won, Calgary-based money manager Josef Schachter was asked to name his top stock pick for the year. Schachter pointed to Golden Rule. "In the next two or three months," he said, "there will be a lot of drilling and trenching occurring [at Stenpad], which could have the ability of proving up anywhere between 20 to 40 million ounces of gold." Up went the stock, from $8.40, to $9.65, to $12.40.

Schachter later admitted that his comments may have been hasty, but he refused to recant. In a follow-up interview in *The Globe and Mail*, Schachter said that "people are saying yes, it's too early to know, but the area has a rich history and therefore, the likelihood of there being no depth to the deposit is not high." In another interview with *The Financial Post*, Schachter predicted that Golden Rule stock "could be worth north of $50, maybe even $100."

He wasn't the only expert picking Golden Rule. The company met with a group of Toronto-based analysts following the publication of Schachter's remarks and referred to a new set of soil samples collected at Stenpad. These indicated even higher grades of gold, as much as ten grams per tonne, over an area of two square kilometres. Jim Steel, an analyst with Newbridge Capital Inc., left the meeting convinced. "It's big, it's real, and it's there," he told reporters.

By dangling large gold-recovery figures in front of a crowd of analysts, Golden Rule officials, led by president Glen Harper, were obviously promoting their company. Yet they refused to acknowledge any responsibility for the concurrent rise in Golden Rule stock. When asked by regulators to explain the jump in price and in trading volume, the company referred to "positive reports from mining analysts." The company attempted to disassociate itself from the Street buzz, noting that "it is very early in the exploration process and premature to attempt to speculate on the size and the grade of the gold resource on the Stenpad property." But what little technical data was out there — and it was spectacular — had of course been supplied by Golden Rule.

The TSE allowed trading to continue, even though further scrutiny led to more doubts. Getting to the bottom of Golden Rule became a giant guessing game. Analysts admitted they had been prevented from visiting the site. By mid-February, the stock had fallen below the $10 mark. As the world shifted its focus back to Bre-X and the unfolding debacle at Busang, Golden Rule discovered that its surface ore samples were — big surprise — bogus. Independent testing revealed the actual gold content at surface was a tiny fraction of what Golden Rule had previously told analysts. On 16 May, two weeks after the final Bre-X meltdown, Harper revealed the discrepancy, while denying his company bore any responsibility. He suggested a "Ghanaian geologist" may have wanted "to keep us happy by putting a whole bunch of quartz that he knew ran [with gold] in our original samples. That is a probability or a possibility. I just don't know."

Unlike Bre-X and Delgratia, Golden Rule lives on, barely. The company made enough cautionary statements and took pro-active steps during the promotion to satisfy stock market regulators. Its stock has never recovered, however, and trades in the $1 range. Glen Harper survived the ordeal; in fact, he profited handsomely, making over $800,000 in 1997, through a combination of salary, option sales and other "compensation." Golden Rule continues to trade on the TSE, although it's now associated with manipulation. Despite the lack of hard proof that its samples were actually salted, the fact remains that investors were badly burned because Stenpad was advertised as something it wasn't. And that should never have happened.

There are serious problems within the junior mining industry, and efforts to address them should be welcomed. A good start would be to view all defenders of the status quo with suspicion. The directors of Canada's four largest securities regulators have already concluded that the Bre-X fraud was "not the result of any systematic flaw in the country's regulatory regime." They claim it was an isolated case. But it wasn't. Bre-X was the worst case of mining fraud in our time, but it wasn't the only example.

That's not to say there aren't any honest junior mining companies scouring the globe for new opportunities. Yet the remarkable
bull market that propelled penny stocks into the big leagues ended
with Bre-X; raising money in the mining sector became virtually
impossible. In 1997, thirty per cent of Canada's biggest stock market
losers were junior mining companies. The TSE's gold and precious
metals subindex, which lists the sector's "best" performers, dropped
almost forty per cent in the same year, the worst showing among all
fourteen of the TSE's subgroups. By contrast, the TSE's financial services subindex, which includes the nation's largest major banks and
brokerages, was up nearly fifty-two per cent.

Falling gold prices were responsible for much of the industry's
misfortune, and senior gold producers were hit especially hard. But
the Bre-X debacle cast doubt and cynicism over the entire Canadian
mining industry. Sadly, the bloodbath that Bre-X precipitated could
have been avoided had a few simple regulatory requirements been
in place.

Reporting requirements must be standardized and made more
rigorous. There are presently ten securities commissions in Canada,
all with different sets of reporting and disclosure codes. By their own
admission, they don't have the resources needed to effectively regulate capital markets. Some industry watchers argue a national securities commission is needed instead of regional ones, and that it must
be given powers to identify, remove and penalize fraud artists and
stock market cheats. Under the current system, criminal actions are
rare; most commissions choose to pursue quick, cost-effective settle-
ments, which usually leave troublemakers free to operate in other
Canadian jurisdictions.

Accountability plays second fiddle to convenience. In Australia,
however, where there is one national stock exchange, and one
national securities regulator, mining companies must follow a single
set of rules or face prosecution. A so-called "competent person" is
responsible for preparing resource and reserve estimates and can be
held accountable if their figures don't jibe with reality.

Further steps could be taken that would virtually guarantee

against mining fraud. Stock market regulators could insist that all geological work be subject to independent audits, conducted at the source, before it is made public. Unfortunately, this isn't likely to happen. Although appropriate technical audits would remove the debilitating stink of fraud from the mining sector, industry insiders argue they would be too costly and time-consuming. Canada's largest stock exchange, the TSE, has chosen not to implement them. Following the Bre-X and Golden Rule debacles, the TSE announced instead that mining companies under its purview must be able to provide information on drilling methods and sampling and assaying procedures.

This new requirement is utterly useless. The Bre-X case is proof of that. Bre-X shared its testing procedures with everyone. That didn't stop someone at the company from salting the drill core, nor did it prevent Bre-X and the country's top-ranked gold analysts from completely overstating the mineral resources at Busang.

In the absence of independent technical audits, it becomes imperative that mining promoters take full responsibility for their statements, instead of blaming employees or associates when something goes wrong. Once again, the TSE has a chance to make solid improvements in the area of corporate disclosure. In fact, the exchange has already commissioned an internal report dealing with the topic. Under the present circumstances, the report notes, "existing remedies available to an investor who suffers losses due to a misrepresentation are so impractical as to be illusory."

The report's authors recommend that stock promoters assume liability for misrepresentative statements, including press releases, which "might reasonably be expected to affect the market price of a security." They also recommend that company directors, officers, "experts" and "influential persons" share some burden of liability. At the time of this writing, none of the report's recommendations had been implemented.

In the end, the investor's best defence may be simple common sense. If something sounds too good to be true, it probably is. Do not assume that analysts know anything they write about. Discount all advice offered from brokerages that underwrite the same stocks they recommend. Remember that the Street is a dog-eat-dog place. Remember that people lie.

Things change, and they don't. Mining booms work in cycles; metals prices rise, promoters jump on the bandwagon, a few genuine discoveries are made, scams are exposed, and for a while frightened investors look for other places to put their money. The industry lies fallow now, but there will be another round of spectacular discoveries, failures, humiliations. Depressed metals prices have forced most companies to rethink their long-term strategy. In hindsight, Inco's $4.3-billion purchase of the Voisey's Bay nickel deposit in Labrador seems foolish, given plunging nickel prices. Inco also underestimated the resolve of local Innu, who claim the land is part of their traditional hunting ground. The Innu have demanded that Inco reimburse them with substantial royalties and have managed to delay construction of an airstrip at the site.

Barrick struggled in the wake of Bre-X. Plunging gold prices sliced deeply into its bottom line, and its stock dropped thirty-five per cent between February 1997 and February 1998. In an effort to convince the world's leading industrialized nations to not eschew gold and liquidate their bullion stockpiles, an increasingly popular practice during periods of low inflation, Brian Mulroney was dispatched to Europe, where he lobbied central bankers on Barrick's behalf. In September 1997, after announcing it was looking at fresh mining opportunities in other distant locations, Barrick flew Mulroney to Irkust, Siberia, to win support from local authorities. At the same time, the company all but gave up on Indonesia, formally withdrawing from two exploration concessions. At least thirty-one other Canadian exploration outfits quickly followed suit.

Only fifteen of approximately fifty Canadian companies that had applied for seventh-generation Contracts of Work stuck around to see their proposals ratified in February 1998. The applications had been delayed for months, after Ida Bagus Sudjana, Indonesia's

avaricious mining minister, tried to "renegotiate" the initial agreements and increase his government's share of every deposit. When
that failed, Sudjana demanded that each applicant pay his department a $10,000 fee, to cover unspecified "photocopying expenses."
For most Canadian companies, this was the last straw, but some
decided to stick it out. "We paid [the $10,000 fee]," admitted
Harold Jones, chief geologist for Pacific Amber. "We decided to play
along." So did officials with Dayak Gold. "It looked like we were
being asked to contribute to a political slush fund," the company's
executive vice-president, William Burton, told me. "But we paid it
in order to see things through." They were betting that Sudjana
would finally be removed from office in March 1998, following President's Suharto's imminent re-election and cabinet shuffle, and they
were right. Sudjana's replacement: Kuntoro Mangkusubroto, his
Machiavellian subordinate. "This is a surprise," Kuntoro said, after
his appointment was announced. "I did not dream of becoming
mines and energy minister." And pigs fly.

Things are quiet now at the Smuggler's Arms, the south Jakarta
bar where hundreds of expatriate miners used to gather to drink and
swap stories. The good life is over; many of bar's regular patrons
have disappeared and left the country, perhaps forever. Indonesia's
economy has collapsed, thanks to a breathtaking plunge in the
country's monetary unit, the rupiah. By the end of 1997, Indonesian
companies owed US$74 billion to outside lenders. This preceded a
series of loan defaults and bank failures. Despite a US$40-billion res-
cue package from the International Monetary Fund, the future
remained bleak. Inflation was rampant; inevitably, food shortages
were followed by horrifying reports of looting and rioting.

Suharto's New Order is bankrupt. The president can no longer
even boast that his country is self-sufficient in rice. Despite promises to dismantle the country's irksome monopolies and reform its
banking system, there are indications that Suharto, now in his
fourth decade of power as national leader, will cling to his old socio-
economic power structure. It's broken, but he knows no other way.
Multinational companies are now unwilling to invest in the country.

In some respects, the mining industry's highly publicized retreat has only compounded Indonesia's problems.

Perhaps the only citizens to benefit from the miners' flight are the most deserving ones, Indonesia's indigenous people, who have been spared the bulldozers and poisonous tailings and blind foreign greed, at least for a while. I'm thinking of the villagers of Long Tesak, whose proximity to Busang had made them prime candidates for forced relocation. I remember the dark river running fast behind Ngang Bilung's house, and slicing into pineapples plucked fresh from his garden, children playing tag with their little brown dogs, squealing as their mothers slowly walked in from the fields, baskets full and heavy on their backs.

A bell clanged, and the spindly old men settled beneath a single bare light bulb, offering their guests sticky black tobacco and fizzy rice beer. How do you feel, I asked Ngang Bilung, now that the Canadians are gone?

"Happy and sad," he said. "There was a lot of embarrassment, people acting badly. But things are back to normal. It's for the best."

BIBLIOGRAPHY

Newspapers

The Australian
Australian Financial Review
Calgary Herald
Globe and Mail
Financial Post
Indonesian Observer
Indonesian Times
Jakarta Post
Melbourne Age
New York Times
Ottawa Citizen
Toronto Star
Vancouver Sun
Wall Street Journal

Reports

ANZ McCaughan Securities Limited. *Asian Mining Review: An Australian Perspective.* July 1995.
Australian Council for Overseas Aid. *Trouble at Freeport: Eyewitness Accounts of West Papuan Resistance to the Freeport-McMoRan Mine in Irian Jaya, Indonesia.* April 1995.

Commission of Inquiry into Discharge of Cyanide and Other Noxious
   Substances into the Omai and Essequibo Rivers. *Report.* January 1996.
Guyana Geology and Mines Commission. *Final Report on Technical Causation,
   Omai Tailings Dam Failure.* January 1996.
Forensic Investigative Associates Inc. *FIA Interim Report of Investigation into
   Tampering with Bre-X Minerals Ltd. Busang Core Samples.* October 1997.
Institute of Policy Research and Advocacy. *Human Rights and the Mining Industry
   in Indonesia: Summary of Research Findings.* February 1997.
Investment Dealers Association of Canada. *Strengthening Emerging Company
   Markets for Investors and Business: Interim Report.* February 1997.
Strathcona Mineral Services Limited. *Busang Project: Technical Audit for Bre-X
   Minerals Ltd.* May 1997.
Toronto Stock Exchange. *Responsible Corporate Disclosure: A Search for Balance.*
   March 1997.
United States of America Department of State. *Indonesia Report on Human Rights
   Practices for 1996.* January 1997.
Report of the Catholic Church. *Violations of Human Rights in the Timika Area of
   Irian Jaya.* October 1995.

Books and Journals

Agricola, Georgius. *De Re Metallica..* Dover Publications, Inc., New York, 1950.
Baines, David. 'Your Risk, His Reward', *Canadian Business*, June 1997.
Baksh, Nazim. 'The Guyana Gold Mining Disaster: Poison in the Lifeline', *CAQ*,
   Spring, 1996.
Croft, Roger. *Swindle! A Decade of Canadian Stock Frauds.* Gage Publishing Ltd.,
   1975.
The Economist. 'If Indonesia Erupts', August 3, 1996
The Economist. 'Suharto's End Game', July 26, 1997
Eliot, Joshua. *Indonesia Handbook.* Footprint Handbooks Limited. Bath, 1996.
Fetherling, Douglas. *The Gold Crusades: A Social History of Gold Rushes.*
   Macmillan of Canada, Toronto, 1988.
Hutchinson, Brian. 'The Prize', *Canadian Business*, March 1997.
Hutchinson, Brian. 'Cigar-Chomping Pastor and His Oilfield Scam', in *The
   Boom and the Bust: 1910-1914* . United Western Communications Ltd.,
   Edmonton, 1994.
Jackson, Richard. *Ok Tedi: The Pot of Gold.* The University of Papua New
   Guinea, Boroko, 1986.
Mackay, Charles. *Extraordinary Delusions and the Madness of Crowds.* Richard
   Bentley, London, 1841.

McBeth, John. "The Battle for Busang," *Far Eastern Economic Review*, December 19, 1996.

Mealey, George. *Grasberg.* Freeport-McMoRan Copper & Gold Inc. New Orleans, 1996.

Miskelly, Norman. 'International Standards for Resources Reporting: Australia's Leadership Role', *ASX Perspective,* July 1997.

Morgenson, Gretchen. 'See No Evil, Speak No Evil', *Forbes*, December 15, 1997.

Pease, Lisa. 'Indonesia, President Kennedy, and Freeport Sulphur', Probe, March/April 1996.

Queenan, Joe. 'Scam Capital of the World', *Forbes*, May 29, 1989.

Ross, Alexander. 'The Resurrection of Windfall's Fallen Woman', *Canadian Business*, December 1991.

Rumball, Donald. *Peter Munk: The Making of a Modern Tycoon.* Stoddart Publishing Co. Ltd., Toronto, 1996.

Schwarz, Adam. *A Nation in Waiting: Indonesia in the 1990s.* Westview Press, Boulder, 1994.

Schwartz, Adam. 'Indonesia after Suharto', Foreign Affairs, July/August 1997.

Shaffer, Ivan. *The Stock Promotion Business: The Inside Story of Canadian Mining Deals and the People Behind Them.* McClelland and Stewart Limited, Toronto, 1967.

Simatupang, Marangin; Sigit, Soetaryo; Wahju, Beni; editors. *Mining Indonesia: Fifty Years Development 1945-1995.* Indonesian Mining Association, Jakarta, 1996.

Suharto. *My Thoughts, Words, and Deeds: An Autobiography* (English edition). PT. Citra Lamtoro Gung Presada, Jakarta, 1991.

Sudarsono; Budi, Johan. 'Adnan's Shadow Behind Sudjana', *Forum Keadilan,* January 13, 1997

Sudarsono; Tjiauw, Sen; Muryadi, Wahyu; Hasyim, Tony; Budi, Johan; Imanullah, Fahmi. 'Politically Imbued Fight for Fortune', *Forum Keadilan,* January 13, 1997.

Tadich, Alexander. *Rampaging Bulls: Outfox Promoters at their Own Game on any Penny Stock.* Elan Publishing Inc., Calgary, 1992.

Wilson, Forbes. *The Conquest of Copper Mountain..* Atheneum, New York, 1981.

INDEX

ABOUT THE AUTHOR

BRIAN HUTCHINSON has written for *Canadian Business* magazine, *Saturday Night* magazine, *The Financial Post, The Financial Times,* and *Alberta Report.* He has been nominated four times for National Magazine Awards. A native of Calgary, he has covered the Canadian business scene for the last nine years. This is his first book.